# Communications in Asteroseismology

Volume 148
December, 2006

Austrian Academy of Sciences Press
Vienna 2006

Editor: **Michel Breger**, Türkenschanzstraße 17, A - 1180 Wien, Austria
michel.breger@univie.ac.at
Layout: **Paul Beck**
beck@astro.univie.ac.at

http://www.univie.ac.at/tops/

COVER ILLUSTRATION
Color-Magnitude Diagram for NGC 2660 with a special focus on variable stars.

British Library Cataloguing in Publication data.
A Catalogue record for this book is available from the British Library.


ISBN 13: 978-3-7001-3805-1
ISBN 10: 3-7001-3805-9
ISSN 1021-2043


Austrian Academy of Sciences Press
A-1011 Wien, Postfach 471, Postgasse 7/4
Tel. +43-1-515 81/DW 3402-3406, +43-1-512 9050
Fax +43-1-515 81/DW 3400
http://verlag.oeaw.ac.at, e-mail: verlag@oeaw.ac.at

# Contents

*Comm. in Asteroseismology*
*Vol. 148, 2006*

# Introductory Remarks

We hope that you will find vol. 148 interesting and scientifically stimulating. CoAst is evolving: the number of satellite astronomy papers has been increasing. This probably reflects a general trend. (Nevertheless, us earthbound astronomers are still in the majority.)

The papers published by CoAst are not only listed in ADS (Astrophysics Data System at NASA), but will now also be included in the SIMBAD database. All papers are refereed - my thanks to all the hardworking referees and the fast responses to these referees' reports by the authors.

The Workshop on the 'Future of Asteroseismology', which was held in Vienna, has just finished. On a personal note, I would like to take the opportunity to thank everybody for their good wishes during the workshop on the occasion of a round birthday. Probably the best gift was the high quality of all the talks: the meeting was a packed powerhouse of information and discussion. Not in the sense of a crowded schedule of talks - in fact, the schedule was quite good allowing for many discussions. 'Packed' in the sense of scientific information that one 'wanted' to listen to. You will be able to read about it in a next issue of CoAst, which will incorporate the proceedings of the workshop.

For many years, Wolfgang Zima has done a wonderful job in helping with the publication of CoAst in Vienna. He has now finished his studies and is a postdoc at Leuven: our loss is their gain. Paul Beck has taken over Wolfi's job.

Michel Breger
Editor

New Link to download Period04 for Windows, Linux, Mac OSX (Intel or Power-PC): http://www.univie.ac.at/tops/Period04/

*Comm. in Asteroseismology*
*Vol. 148, 2006*

# The Potential of Non-standard Pulsation Models for Asteroseismology

T. Collins[1], J. Guzik[1]

[1] Los Alamos National Laboratory

## Abstract

The process of building seismic models of observed stars that match the observed pulsation frequencies often proves to be a difficult one. The outputs of computational codes used to perform these calculations match observational data fairly well for the sun, but fail to do so as well for larger-mass stars. This suggests that the codes are neglecting potentially significant processes that occur more strongly in larger stars. We examine some hydrodynamically motivated modifications to the standard stellar models that might help to bring theoretical calculations into agreement with observational data. The analysis of Press (1981) and Press & Rybicki (1981) is applied to the case of the $\delta$ Scuti star FG Vir, and presented.

## Introduction

Current stellar evolution and pulsation models rely on a number of simplifying assumptions in order to make the problem of stellar modeling more tractable, especially when dealing with the hydrodynamics of stellar interiors. Stars exhibit a wide variety of hydrodynamic processes that affect the evolution and structure of a star to some extent, while most stellar evolution and pulsation models neglect the majority of these effects in favor of computational efficiency and simplicity. For example, the mixing-length treatment of Böhm-Vitense (1958) still seems to be the standard in general convection treatments. Non-convective instabilities are often ignored altogether in stellar evolution modeling.

This set of hydrodynamic approximations has worked quite well for the solar case. Despite the recently increased discrepancy between solar models and helioseismology with the new solar abundances these assumptions still yield far better results for the sun than they do for $\delta$ Scuti stars. Previous attempts at modeling the $\delta$ Scuti star FG Vir by Breger et al. (1999), Templeton et al. (2001), and Guzik et al. (2004) have met with difficulty and have yet to produce a model that can match all the frequencies observed. Templeton et al. (2001) and Guzik et al. (2004) however both had some success in improving the agreement between the predicted and observed frequencies via increased mixing between the convective core and the stable envelope. In the case of Guzik et al. (2004) core mixing was introduced into the model, not through the use of a convective overshooting parameter, as was done by Templeton et al. (2001), but instead by applying a large opacity multiplier from the core out into the radiative transport region. This multiplier caused approximately $0.8 - 0.9\ H_p$ overshooting. However Deupree (2000) found in 2-D hydrodynamic simulations that for stars with under 3 $M_\odot$ overshooting was only found to extend to $0.3\ H_p$. Even for more massive stars Deupree (2000) found that mixing extended only to a height of $0.45\ H_p$. The amount of overshooting induced by Guzik et al. (2004) was probably too high; however the level of mixing might be justified as there are other processes that could mix hydrogen-rich envelope material into the core.

In the solar case, agreement between theoretical and observed pulsation frequencies and sound speed profiles was worsened by a downward revision of the solar metallicities. This has prompted a large number of papers examining a range of possible modifications to the solar models that could restore the previous agreement. Bahcall et al. (2005) found that a small opacity increase of roughly 10% between $2 \times 10^6 K$ and $5 \times 10^6 K$ (the temperature regime in the region below the convective zone) very nearly restored the agreement that had been lost in the sound speed profiles, predicted surface helium abundance, and the depth of the convective zone.

In the two separate cases of both the Sun and FG Vir opacity modifications near the boundaries of convective regions inserted into otherwise standard evolution and pulsation codes have resulted in improved agreement between predictions and observations. This suggests the possibility of a potentially significant hydrodynamic effect in the stable layers near the boundaries of the convective regions. This possibility was brought to our attention by Arnett, Meakin, and Young (private communication, in preparation). With this possibility in mind we have undertaken a large survey of possible models of FG Vir examining modifications near the surface convective zone and the core convective zone, both separately and in combination to see if this possibility could improve the agreement between frequency predictions and observation for this object. The results of that computational study are to be presented in Collins & Guzik (in preparation). Our focus here is to consider the theoretical justification for that study.

## Opacity Modification by $g$ modes Excited by Convective Overshooting

We consider now the work of Press (1981) and Press & Rybicki (1981) (from here on P81 and PR81 respectively), who carefully examined exactly this possibility in detail. We begin with considering the characterization of the internal wave. P81 chose to represent the fluid flow of an internal wave in Cartesian coordinates to be

$$v_x = -u \cos\alpha \sin(k_V z + k_H x + \omega t) \quad (1)$$
$$v_y = 0 \quad (2)$$
$$v_z = u \sin\alpha \sin(k_V z + k_H x + \omega t) \quad (3)$$

We wish to show that the results of their analysis are invariant under rotations of $\vec{k}$ with the constraint that $k^2 = k_V^2 + k_H^2$. The motivation for this being that in higher mass stars there are multiple convective regions to excite propagating g modes that would propagate both inward and outward. If the effective opacity increase predicted by PR81 is independent of the direction of wave propagation, then it should be more broadly applicable than the limited solar case considered in P81.

As in PR81 we apply the local Boussinesq approximation so that $\nabla \cdot \vec{v} = 0$ which gives

$$\tan\alpha = \frac{k_H}{k_V} \quad (4)$$

so that $\alpha$ is the angle between the wave vector $\vec{k}$ and the z-axis. As described by P81 we can estimate $\alpha$ if we can estimate $\omega$. We can do this by using the results for $k_V$ and $k_H$ given by P81

$$k_H \equiv \left[\frac{l(l+1)}{r^2}\right]^{1/2} \quad (5)$$

$$k_V \equiv k_H \left[\frac{N^2}{\omega^2} - 1\right]^{1/2} \quad (6)$$

We can estimate $\omega$ from the convective turn over frequency.

We apply a coordinate transformation so that the wave vector lies along one axis.

$$\begin{bmatrix} \chi \\ \gamma \\ \zeta \end{bmatrix} = \begin{bmatrix} \cos\alpha & 0 & -\sin\alpha \\ 0 & 0 & 0 \\ \sin\alpha & 0 & \cos\alpha \end{bmatrix} \begin{bmatrix} x \\ y \\ z \end{bmatrix} \quad (7)$$

We transform equations (1-3) into this coordinate system where

$$v_\chi = -u\,sin\,(k\zeta + \omega t) \tag{8}$$

$$v_\gamma = 0 \tag{9}$$

$$v_\zeta = 0 \tag{10}$$

Thus for an internal wave flow the actual fluid flow is always perpendicular to the wave vector.

The effective modification to the opacity from P81 is given by

$$< \kappa_{ij} >= \frac{\kappa_0}{4\omega^2}\left(\sigma_{ik}\sigma_{kj} + 2\sigma_{ik}\sigma_{jk} + \sigma_{ki}\sigma_{jk}\right) \tag{11}$$

where

$$\sigma_{ij} = \frac{\partial v_i}{\partial x_j} \tag{12}$$

which gives

$$\sigma_{ij} = \begin{pmatrix} 0 & 0 & -uk \\ 0 & 0 & 0 \\ 0 & 0 & 0 \end{pmatrix} \tag{13}$$

$$< \kappa_{ij} >= \kappa_0 \frac{u^2k^2}{2\omega^2} \begin{pmatrix} 1 & 0 & 0 \\ 0 & 0 & 0 \\ 0 & 0 & 0 \end{pmatrix} \tag{14}$$

The heat flux is shown by PR81 to be

$$F_i = -\kappa_0 \frac{\partial T}{\partial x_i} - \kappa_{ij}\left(\frac{\partial T}{\partial S}\right)_P \frac{\partial S}{\partial x_j} \tag{15}$$

which is equivalent to

$$F_i = -\left[\kappa_0\delta_{ij} + \left(1 - \frac{\nabla_{ad}}{\nabla}\right)\kappa_{ij}\right]\frac{\partial T}{\partial x_j} \tag{16}$$

where $\nabla_{ad}$ and $\nabla$ have the usual meanings. Applying (14) to (16) gives

$$\begin{bmatrix} F_\chi \\ \\ F_\gamma \\ \\ F_\zeta \end{bmatrix} = \begin{bmatrix} -\kappa_0\left(1 + \left(1 - \frac{\nabla_a d}{\nabla}\right)\frac{u^2k^2}{2\omega^2}\right)\frac{\partial T}{\partial \chi} \\ \\ -\kappa_0 \frac{\partial T}{\partial \gamma} \\ \\ -\kappa_0 \frac{\partial T}{\partial \zeta} \end{bmatrix} \tag{17}$$

Thus, the radiative flux is modified only along the axis of fluid flow($\chi$-axis), perpendicular to the direction of wave propagation ($\zeta$-axis). If we assume that the temperature gradient is aligned with the vertical z-axis (an assumption inherent to one dimensional stellar evolution codes) then via the use of (7) it is easily shown that

$$\frac{\partial T}{\partial \chi} = -sin\ \alpha \frac{\partial T}{\partial z} \tag{18}$$

This implies of course that the larger $k_V$ is relative to $k_H$, the weaker the modification of the radiative flux. Thus as a consequence of (6), higher frequency waves will more strongly alter the radiative opacity. The same is true of waves with higher $l$ values, as a consequence of (5).

The key conclusion that follows from (17) is that internal waves always act to reduce the radiative flux, regardless of the orientation of $\vec{k}$ relative to $\nabla T$. This implies that the effective opacity increase cannot be related to the inward luminosity carried by the energy flux of the propagating internal waves, because reversing $\vec{k}$ does not alter the effect. This conclusion is important for the asteroseismology of larger stars, because this effect should be seen in the stable interior of the star near both the core convective zone, and the envelope convective zone.

## Enhanced Mixing by $g$ modes Excited by Convective Overshooting

We briefly consider in a similar manner the results of PR81 for the enhanced diffusion of a passive contaminant. In the case of heat transport above, it is the diffusion of entropy that is enhanced by the fluid motion, which results in the effective optical opacity increase. Similarly, the diffusion of particles is enhanced by fluid motion where

$$F_i = -\left[\kappa_0 \delta_{ij} + < \kappa_{ij} >\right] \frac{\partial \theta}{\partial x_j} \tag{19}$$

where $\theta$ is the concentration of some particle species. As above it follows that

$$\begin{bmatrix} F_\chi \\ \\ F_\gamma \\ \\ F_\zeta \end{bmatrix} = \begin{bmatrix} -\kappa_0 \left(1 + \frac{u^2 k^2}{2\omega^2}\right) \frac{\partial \theta}{\partial \chi} \\ \\ -\kappa_0 \frac{\partial \theta}{\partial \gamma} \\ \\ -\kappa_0 \frac{\partial \theta}{\partial \zeta} \end{bmatrix} \tag{20}$$

where particle diffusion is enchanced along the axis of fluid motion.

In addition to the enhancement of particle diffusion, large amplitude internal waves can become unstable to subwavelength turbulence (PR81). The resulting mixing can extend the region of homogenous composition beyond the boundaries of the convective regions approximately out to the linear damping distance of the excited $g$ modes (P81).

## Conclusions and Future Work

The most important conclusion is that stochastically excited $g$ modes that are expected to be found at the interface between convective and stable regions have the same qualitative effect of diffusion enhancement as described by PR81, regardless of the relative orientation between the wave vector and the diffusive flux being considered (represented by the angle $\alpha$)–only the strength of the effect is dependent on alpha, not its qualitative nature.

This has important conclusions for the inclusion of this phenomenon into standard stellar models. In the case of radiation transport we expect to see enhanced diffusion of entropy which damps radiative transport. We expect to see this whether the $g$ mode causing the effect is propagating inward or outward, that is to say that the effect is independent of kinetic energy transported by the $g$ mode. Therefore it should be possible to model the effect of internal wave flow on radiative transport with a modification of optical opacities near the interfaces of all convective and stable regions. Further, we expect the region of homogenous composition around these cores to be enlarged beyond what would be expected from the overshoot heights found by Deupree (2000), especially important for the cases of larger mass stars.

In addition to the supression of radiative transport near the interfaces of convective and stable regions, we also expect to see a strengthening of particle diffusion. Keep in mind that the effect of the internal wave flow is to enhance all diffusion by some scaling factor along the direction of fluid flow. Gravitational diffusive settling should therefore be increased by this mechanism.

It would be worthwhile to produce a general theoretical formalism for expressing this effect that could be implemented in existing stellar evolution codes. Ideally such a formalism would make no assumptions about the particular convection treatment chosen and instead characterized by parameters such as the convective turnover frequency or mean flow velocity that would be predicted by the convective treatment.

## References

Bahcall J.N., Basu S., Pinsonneault M., & Serenelli A.M. 2005, ApJ, 618, 1049
Breger M., Handler G., Nather R.E., et al. 1995, A&A, 297, 473
Breger, M., Pamyatnykh, A. A., Pikall, H., & Garrido, R. 1999, A &A, 341, 151
Böhm-Vitense, E. 1958, Z. Astrophys., 46, 108

Deupree, R. G. 2000, ApJ, 543, 395 Guzik J.A., Austin B.A., Bradley P.A., & Cox A.N. 2004, In: Kurtz D.W., Pollard
K.R. (eds.) ASP Conf. Ser. 310: IAU Colloq. 193: Variable Stars in the Local Group, 462
Press, W.H. 1981, ApJ, 245, 286
Press, W.H., & Rybicky, G.B. 1981, ApJ, 248, 751
Templeton M., Basu S., & Demarque P. 2001, ApJ, 563, 999

*Comm. in Asteroseismology*
*Vol. 148, 2006*

# NGC 2660 revisited

T.E. Mølholt[1], S. Frandsen[1], F. Grundahll[1,2] and L. Glowienka[1]

[1] Dept. of Physics and Astronomy, University of Aarhus, Ny Munkegade, Bygn. 1520, DK 8000 Aarhus C, Denmark
[2] Danish AsteroSeismology Centre, University of Aarhus, Ny Munkegade, Bygn. 1520, DK 8000 Aarhus C, Denmark

## Abstract

A new search for variable stars in NGC 2660 has revealed a number of interesting targets. The cluster has more $\delta$ Scuti members than seen before and an eclipsing binary, where one of the components may be a variable star too.

## Introduction

Open clusters as well as globular clusters are becoming increasingly important for the test of stellar evolution and for studies of individual stars of special importance, where one can use the membership to infer some of the basic properties of the star (age, metal content etc.). Some years ago we initiated a programme to search for $\delta$ Scuti stars in open clusters, which together with other similar searches have produced a list of clusters with many $\delta$ Scuti star members. The philosophy has been, that by analysing many variables at the same time, we could learn more about the individual stars. Due to the very difficult mode identification problem, this has turned out not to provide the answers straight away.

Examples from this programme on clusters are the first observations of NGC 2660 (Frandsen et al. 1989), a campaign on NGC 6134 (Frandsen et al. 1996), variables in NGC 1817 (Arentoft et al. 2005) and NGC 7062 (Freyhammer & Arentoft 2001). NGC 1817 is a particularly interesting case with the largest set of $\delta$ Scuti stars of any cluster. A new cluster added to the set is NGC 2506, which is presently being analysed. Finally, one should also mention the work on nearby clusters (Praesepe and the Pleiades, e.g., Fox Machado et al. 2006).

We would very much like to get the mass of at least one $\delta$ Scuti star by comparing observed frequencies with model frequencies. Then the mass of other stars in the cluster could be inferred from the position on the isochrone in the HR diagram. Unfortunately, we have not been able to match theoretical and observed frequencies very well so far, and no accurate masses have been determined.

Another possibility to get the mass is to locate detached eclipsing binaries in clusters and use these systems to derive accurate masses. Again, we can then use the isochrone to derive masses for other stars, in particular the $\delta$ Scuti stars. This will make the comparison of model frequencies and observed frequencies considerably easier. Using large telescopes/efficient spectrographs the observations are now of a quality, which even for faint stars in clusters lead to accuracies of less than 1% in mass and radius.

## The cluster and the observations

NGC 2660 has an age $t \sim 1$ Gy, a distance module of $(m - M)_0 = 12.2$ corresponding to a distance of $D = 2750$ pc and a reddening $E(B - V) = 0.4$. The metal content is solar: $[Fe/H] \approx 0.0$. The cluster has been observed by Frandsen et al. (1989) and Sandrelli et al. (1999)

During the period Feb. 5 to Feb. 27 2005 new data was obtained with the Danish 1.54m telescope at La Silla during time, where the main targets could not be observed. About 1500 frames in the filters Bessel B (464) and Gunn I (949) were acquired. A small set (45) of V frames were observed as well. The camera has a field of view of 13.7x13.7 square minutes and exposure times were between 60–120 s in B and 20–60 s in I. The V frames are only used to calculate colours and are not part of the time series analysis.

The reduction was done with the DAOPHOT/ALLSTAR (Stetson 1987) code.

## Results

A search for variables in the combined set of old and new data confirms the presence of several $\delta$ Scuti stars and a detached eclipsing binary system close to the cluster Colour-Magnitude turnoff. Some of the variables are indicated in Fig. 1. There is clearly a set of variables above the turnoff, that one would expect to show pulsations on time scales of hours as they are situated in the instability strip.

The variable stars that might be members of the clusters based on the distance to the cluster center and the magnitude are listed in Table 1. The magnitude and colour are only indicative, as they have been derived from different datasets. Therefore no errors are given. The pulsation properties are given in Table 2.

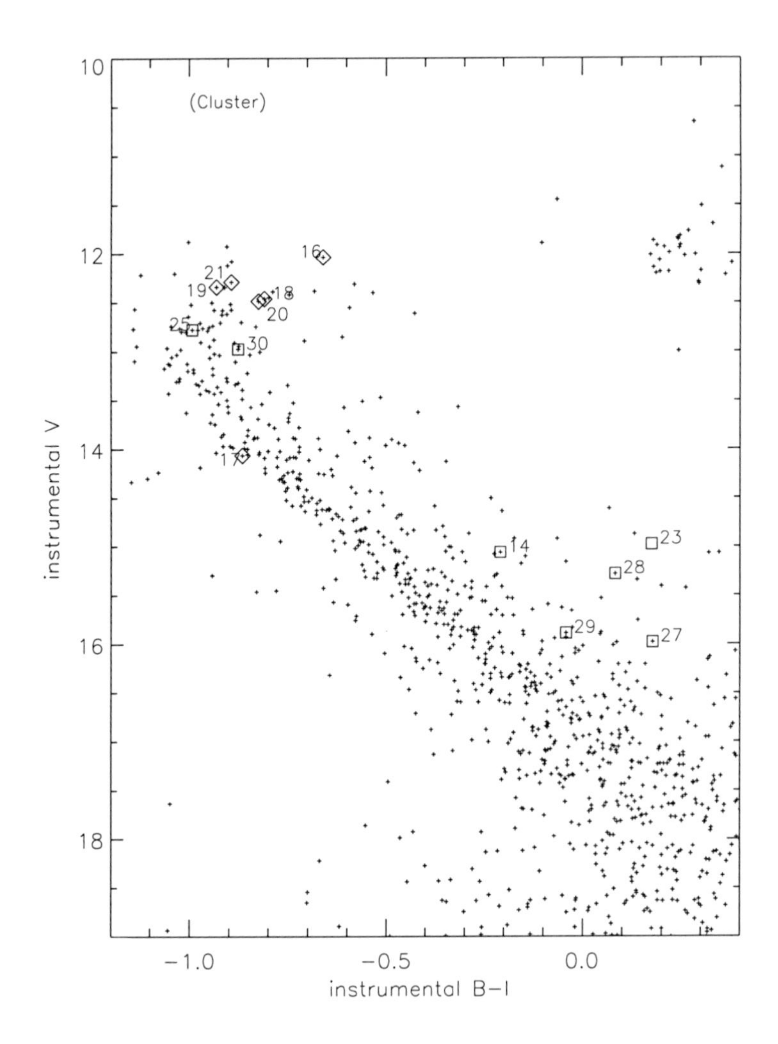


Figure 1: CM diagram for NGC 2660. Diamonds are pulsating stars. Squares are eclipsing binaries or other types of variables.

The important results are the confirmation of the presence of five non-ambiguous $\delta$ Scuti stars, all of them clearly multiperiodic, and the determination of the nature as well as the period 1.46016 d of the detached EB V25.

A few comments and a figure follows for the main variables in the list.

Table 1: **The variables detected in NGC 2660.** *Column 1:* The variable names are defined here. *Column 2:* ID from WEBDA (Paunzen & Mermilliod 2006), *Columns 3 and 4:* Coordinates, *Columns 5-6:* magnitude and colour.

| ID | WEBDA | $\alpha_{2000}$ | $\delta_{2000}$ | V | B-V |
|---|---|---|---|---|---|
| V14 | | 8 42 51.75 | -47 11 43.1 | 17.57 | 1.15 |
| V16 | 552 | 8 42 41.96 | -47 12 09.0 | 14.30 | 0.66 |
| V17 | 637 | 8 42 43.87 | -47 12 31.6 | 16.38 | 0.72 |
| V18 | 289 | 8 42 36.20 | -47 11 29.0 | 14.81 | 0.70 |
| V19 | 672 | 8 42 44.74 | -47 13 01.2 | 14.68 | 0.67 |
| V20 | 395 | 8 42 38.59 | -47 12 06.7 | 14.81 | 0.70 |
| V21 | 174 | 8 42 33.27 | -47 11 34.7 | 14.63 | 0.68 |
| V22 | 323 | 8 42 36.88 | -47 12 27.3 | 18.86 | 1.50 |
| V23 | | 8 42 15.59 | -47 12 27.1 | 18.70 | 1.14 |
| V25 | 518 | 8 42 41.23 | -47 12 06.3 | 15.21 | 0.58 |
| V26 | | 8 42 37.90 | -47 08 41.2 | 18.49 | 1.35 |
| V27 | 205 | 8 42 42.73 | -47 13 35.5 | 18.21 | 1.58 |
| V28 | 589 | 8 42 42.73 | -47 10 35.5 | 17.60 | 1.07 |
| V29 | 411 | 8 42 38.90 | -47 10 37.0 | 18.19 | 1.06 |
| V30 | 406 | 8 42 38.77 | -47 12 38.4 | 15.28 | 0.69 |

### V25

The possible nature of this object was suggested already in Frandsen et al. (1989) based on only 14 datapoints. The new data gives a first precise period but the quality is still not quite sufficient for a precise determination of the stellar parameters. The 'noise' on the flat part of light curve might indicate that one of the components could be variable, but it could also have an instrumental origin.

### The $\delta$ Scuti stars V16, V18, V19, V20 and V21

As easily seen the Fig. 3, 4, 5, 6 and 7 the light curves show the typical period and amplitude changes of pulsating variables with multiple close frequencies. The identified frequencies with $S/N > 4$ are listed in Table 2.

### A contact binary

A possible binary system of two close stars is presented by variable V14. It lies above the main sequence and is considerably fainter than V25.

Table 2: **The variables detected in NGC 2660.** *Column 1:* The variable names are defined here. *Column 2:* ID from WEBDA (Paunzen & Mermilliod 2006), *Columns 3–5:* Property of variability and *Column 6:* comments, where EB=Eclipsing Binary, DS=$\delta$ Scuti star, CM=Cluster Member and GD=$\gamma$ Dor star.

| ID | WEBDA | $f$ c/d | $S/N$ | $a_B$ mmag | Notes |
|---|---|---|---|---|---|
| V14 | | $P = 0.3259$ d | | | EB |
| V16 | 552 | 5.86 | 12.9 | 52 | DS CM |
| | | 6.31 | 10.8 | 44 | |
| | | 10.50 | 4.2 | 17 | |
| V17 | 637 | $P = 0.9$ d | | | GD CM |
| V18 | 289 | 6.48 | 5.9 | 18 | DS CM |
| | | 7.22 | 5.8 | 12 | |
| | | 12.31 | 5.4 | 14 | |
| | | 12.51 | 5.3 | 13 | |
| | | 11.71 | 4.6 | 14 | |
| V19 | 672 | 10.07 | 8.0 | 18 | DS CM |
| | | 11.16 | 5.9 | 14 | |
| | | 10.50 | 5.4 | 12 | |
| | | 4.83 | 5.1 | 13 | |
| V20 | 395 | 7.53 | 6.6 | 26 | DS CM |
| | | 5.91 | 6.5 | 26 | |
| | | 6.19 | 4.2 | 16 | |
| V21 | 174 | 12.78 | 7.5 | 23 | DS CM |
| | | 6.96 | 5.7 | 15 | |
| | | 13.61 | 4.6 | 14 | |
| | | 7.86 | 4.2 | 11 | |
| V22 | 323 | $P = ??$ | | | EB, CM |
| V23 | | $P = 0.389$ d | | | Contact EB |
| V25 | 518 | $P = 1.46016$ d | | | det. EB, CM |
| V26 | | $P = ??$ | | | EB |
| V27 | 205 | $P = ??$ | | | EB |
| V28 | 589 | $P \sim 6$ d | | | Var CM |
| V29 | 411 | $P =$ several d | | | det. EB, CM |
| V30 | 406 | $P = 0.52$ d | | | Var CM |

A possible $\gamma$ Dor star

Variable V17 shows the long period and the location in the CM-diagram of the $\gamma$ Dor variables.

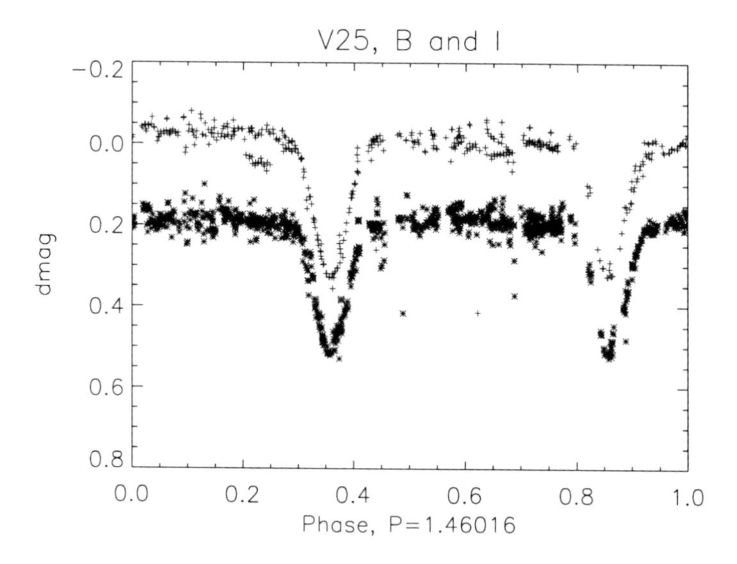


Figure 2: Phase plot for the detached EB V25. The $I$ points have been shifted by 0.2 magnitude. Differential magnitude relative to reference stars are plotted normalised to 0.0 or 0.2 outside eclipses.

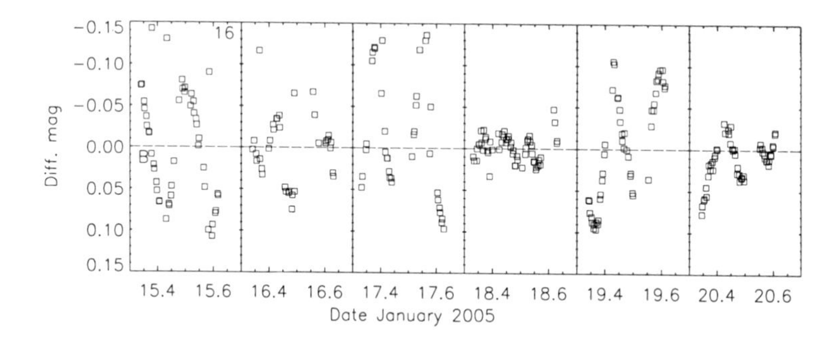


Figure 3: The $B$ light curve for V16. The $I$ data was of poor quality for this star due to crowding.

## Conclusion

NGC 2660 is shown to be one of a small set of open clusters, where one has the potential of determining precise masses for the $\delta$ Scuti stars due to the presence of a detached eclipsing binary system not far away in mass from the pulsating

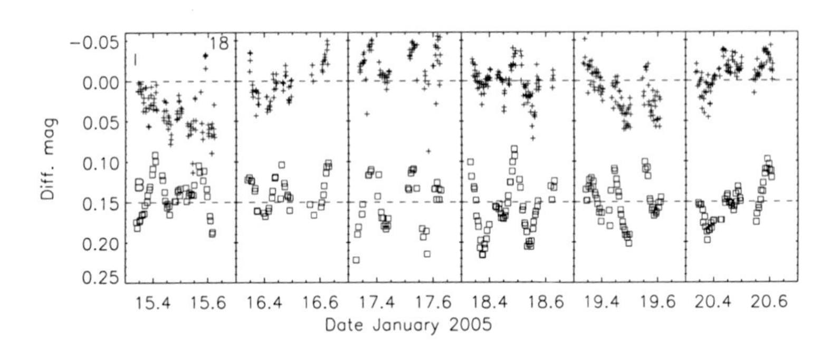


Figure 4: The $B$ and $I$ light curve for V18.

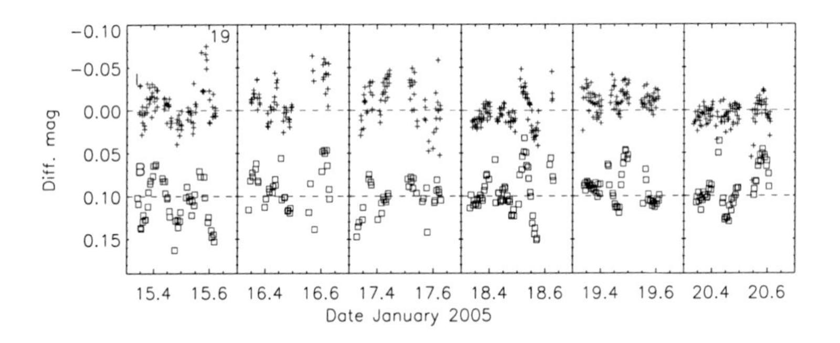


Figure 5: Same for V19.

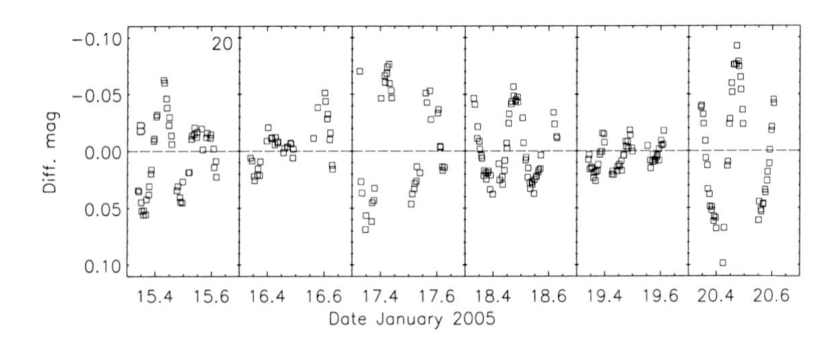


Figure 6: Same for V20, but only B is shown.

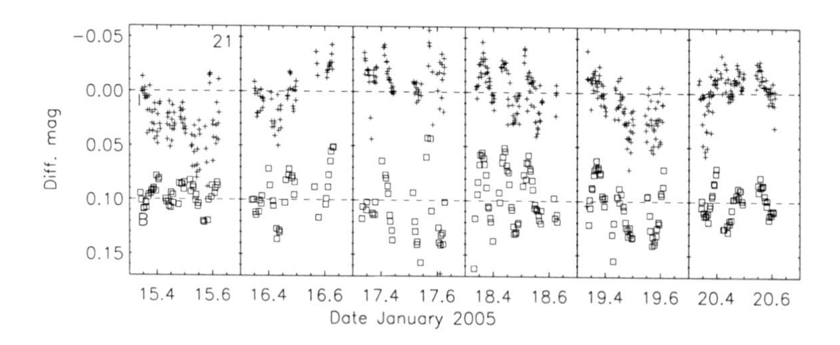


Figure 7: Same for V21.

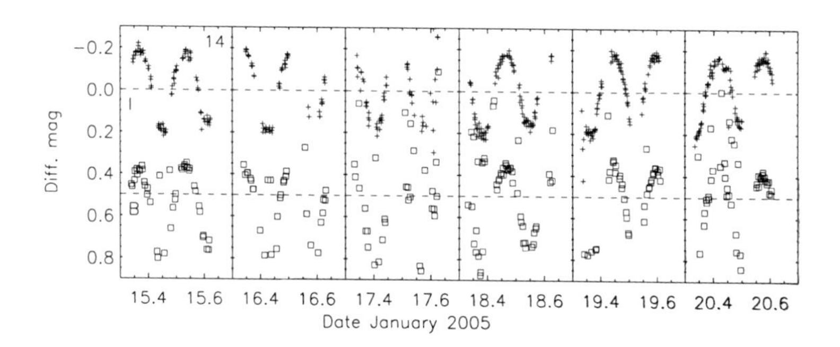


Figure 8: Contact binary V14.

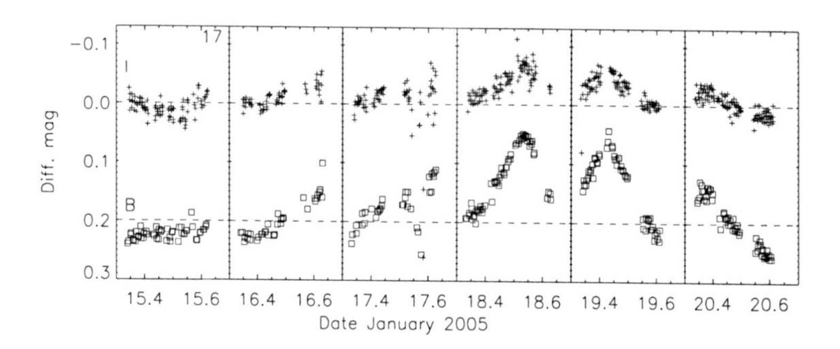


Figure 9: A $\gamma$ Dor variable: 17.

stars. At the same time a better age determination is possible, which also is important when modeling the pulsating stars.

A small but not unimportant result is, that it is possible to find candidates for detached eclipsing binaries from a small set of frames as the confirmation of the nature of V25 shows. This knowledge is useful when searching for detached EBs in other clusters, particularly globular clusters, where detached EBs are rare but extremely important for age determinations.

**Acknowledgments.** Based on observations with the Danish 1.54m telescope at La Silla, ESO. This research was supported by the Danish Natural Science Council through its Instrument Center for Danish Astronomy.

## References

Arentoft, T., Bouzid, M. Y., Sterken, C., et al. 2005, PASP, 117, 601
Freyhammer, L. M., & Arentoft, T., C. 2001, A&A, 368, 580
Frandsen, S., Dreyer, P., & Kjeldsen, H. 1989, A&A, 215, 287
Frandsen, S., Balona, L.A., Viskum, M., et al. 1996, A&A, 308, 132
Fox Machado, L., Pérez Hernandez, F., Suárez, J.C., et al. 2006, A&A, 446, 611
Paunzen, E., & Mermilliod, J.-C. 2006, http://www.univie.ac.at/webda/
Sandrelli, S., Bragaglia, A., Tosi, M., & Marconi, G. 1999, MNRAS, 309, 739
Stetson, P.B. 1987, PASP 99, 121

*Comm. in Asteroseismology*
*Vol. 148, 2006*

# The moving bump in the light curves of SS For and RR Lyr

E. Guggenberger[1], K. Kolenberg[1,2]

[1] Institut für Astronomie, Türkenschanzstrasse 17, 1180 Vienna, Austria
[2] Institute of Astronomy, University of Louvain, Celestijnenlaan 200B, B-3001 Heverlee, Belgium

## Abstract

High-precision multisite photometry was used to investigate the bump occurring before minimum light in the light curves of the two RRab Blazhko stars RR Lyr and SS For. For both stars the phase of the bump was found to be variable with a period equal to the Blazhko period. There seems to be a direct connection between the behavior of the bump and the Blazhko effect. As most models connect the phase of the bump to the stellar radius, a variable bump phase may provide constraints to the models for explaining the bump.

## Introduction

In the majority of RR Lyrae stars a characteristic bump appears in the light curve just before luminosity minimum. Generally, shock or compression waves are thought to be responsible for this phenomenon, as RR Lyr stars don't meet the conditions for resonance that might cause the similar bump in cepheid light curves. Unlike the Hertzsprung progression in classical cepheids, no connection between the phase of the bump and the pulsation period of the star is known. Instead, we notice a strong change in the bump phase during the Blazhko cycle in SS For, one of our target stars observed in the framework of the Blazhko project (see http://www.univie.ac.at/tops/blazhko for project details). The Blazhko effect is a periodic change of pulsation amplitude and/or phase on time scales of usually tens of days (Blazhko 1907). With the luminosity maximum occuring at phase zero, the phase of the bump in SS For varies between 0.65 and 0.85, which is, in other words, 20 per cent of the pulsation period. At the same time the strength of the bump is variable. The bump in RR Lyrae stars is generally believed to be the result of a shock wave, but there are different models to explain the phenomenon. The observed motion of the bump in the phase diagram might help to provide constraints for the models.

## Data and Methods

For our investigation we combined data taken at the South African Astronomical Observatory (SAAO) and the Siding Spring Observatory in Australia (SSO) during a multisite campaign carried out from July through September 2005. SAAO data from the year 2004 as well as data gathered in 2000-2006 by the All Sky Automated Survey (ASAS, Pojmanski 2002) have also been added for analysis. ASAS data are in Johnson *V* filter, SAAO and SSO are both in Johnson *V* and *B*. A publication on the observations and data analysis is being prepared. The strong changes of the light curve shape around minimum light can be seen in Figure 1.

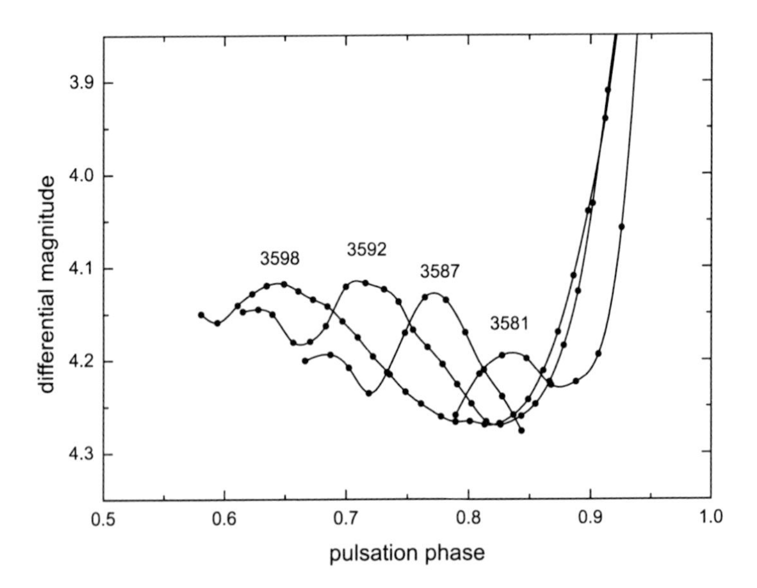


Figure 1: Phase diagram of SS For around bump phase for four different nights. Numbers in the figure represent Julian Date in the form 2450000+. It can be seen how the bump changes in phase and amplitude during time intervals of 5 and 6 days, respectively. Scatter of data points is smaller than the symbols.

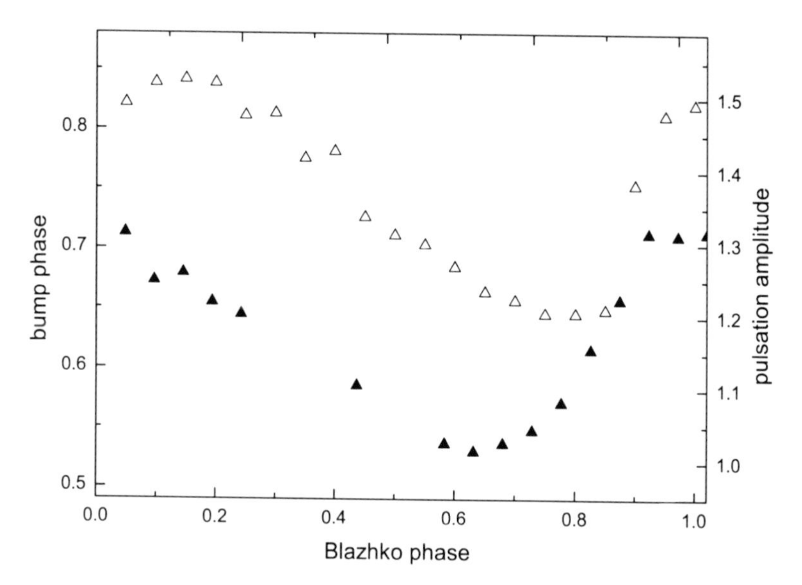


Figure 2: SS For: Change of pulsation amplitude (filled symbols) and phase of the bump maximum (open symbols) during a Blazhko cycle.

From this extensive data set we found a Blazhko period of about 34.7 days for SS For using Period04, a software package for statistical analysis of large astronomical data sets (Lenz & Breger 2005). This period was used to calculate the Blazhko phase for all datapoints and to create phase bins. Using the *V* data, 20 overlapping phase bins of 0.1 of the Blazhko period were created. The low scatter within the phase bins indicates that the correct value for the Blazhko period had been used. For each phase bin the pulsation phase of the bump maximum was measured and plotted against the Blazhko phase. The result can be seen in Figure 2.

We see that the phase of the bump maximum varies between a phase of 0.65 and 0.85, and that the period of this variation is identical to the Blazhko period. Furthermore, it is obvious that the curve of the phase variation is non-sinusoidal, just like the variation of the pulsation amplitude during a Blazhko cycle. It is characterized by a slow progress to lower phases ('to the left') and a quicker motion to higher phases after the Blazhko minimum. The $\varepsilon$-value, which is as a measure of the skewness of a curve defined as

$$\varepsilon = \phi_{\max} - \phi_{\min} \quad (1)$$

is 0.35 both for the variation of the bump phase and the variation of the pulsation amplitude.

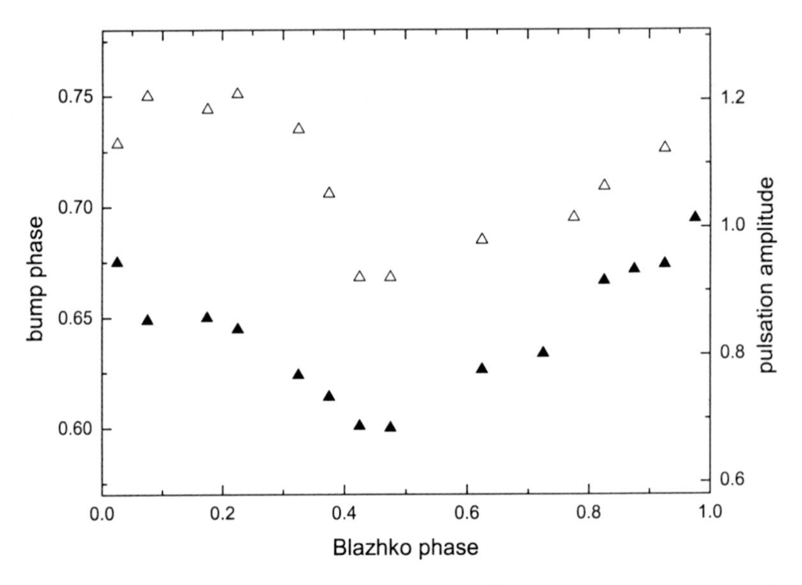


Figure 3: RR Lyr: Change of pulsation amplitude (filled symbols) and phase of the bump maximum (open symbols) during a Blazhko cycle.

The first question arising is whether this variability of bump phase is a common phenomenon among (Blazhko) RR Lyr stars. We performed the same test on a similar data set of RR Lyr. For details on this data set we refer to Kolenberg et al. (2006). In this case the Blazhko period found from the data set was 39 days. Again, the pulsation amplitude and the phase of the bump maximum were determined for each phase bin and plotted with respect to the Blazhko phase. The results turned out to be comparable (see Figure 3), altough the data of RR Lyr are of a lesser quality than those of SS For. Like for SS For, we see that the period of the bump motion is identical to the Blazhko period, and that the minimum bump phase is reached near Blazhko minimum, while the maximum bump phase is reached near maximum pulsation amplitude.
We see that the interval in phase in which the bump occurs is smaller for RR Lyr than it is for SS For. The lowest phase that is reached by the bump maximum in RR Lyr is 0.67, the largest 0.75 with the total light curve maximum set to zero phase. This is a variation of only 8 per cent the pulsation period, compared to the high value of 20 per cent for SS For. The latter star is known for its rather large variations at minimum light (Lub 1977). In most Blazhko stars the variations around maximum light are most pronounced, while those around minimum light are much smaller.

## Models

RR Lyr stars are far from the condition of $P_2/P_0 = 0.5$ that could explain the existence of the bump in Cepheids by resonance. Therefore, shock waves are generally thought to be responsible for the observed bump in the light curves of RR Lyr stars. At present, there are two different models for explaining the bump, both involving shock waves: the 'echo' model and the 'infall' model (Gillet & Crowe 1988). According to the infall model, fast inward moving upper atmospheric layers collide with deeper, slower ones during infall phase (Hill 1972). The echo model, on the other hand, connects the bump to a compression or shock wave that has been reflected on the stellar core.

## Discussion

If the echo model is correct, one would expect a correlation between the stellar radius and the phase of the bump, as the shock wave travels a longer distance in a larger star than in a smaller one. This correlation was examined and confirmed by Carney et al. (1992) who calculated the elapsed time from the preceding minimum radius to the appearence of the bump for different stars and found a relation to the mean radius. This model can only explain a moving bump if the mean or equilibrium radius of the star changes. The infall model, on the other hand, connects the bump phase to the so-called expansion radius, which is given by Carney et al. (1992) as

$$\Delta R = \int p(v_{\rm rad} - \gamma)dt, \qquad (2)$$

where p is the projection factor, $v_{\rm rad}$ is the radial velocity, and $\gamma$ the so-called $\gamma$-velocity during a Blazhko cycle. Though the Blazhko effect is not fully understood yet, the expansion radius is thought to be very likely to change during a Blazhko cycle. This would fit the variation of the phase of the bump as observed in our data. The bump appears late when the star is at Blazhko maximum and therefore is expected to have a larger expansion radius. The bump appears earlier around Blazhko minimum, when the star itself shows a variation that is less strong.

### Distinctiveness of the bump

We also examined the strength of the bump both with respect to Blazhko phase and filter. The bump is stronger in the *B* filter than it is in *V*, as it is clearly visible in a *B*-*V* diagram (see Figure 4). We thus confirm the findings of previous authors such as Gillet & Crowe (1988).

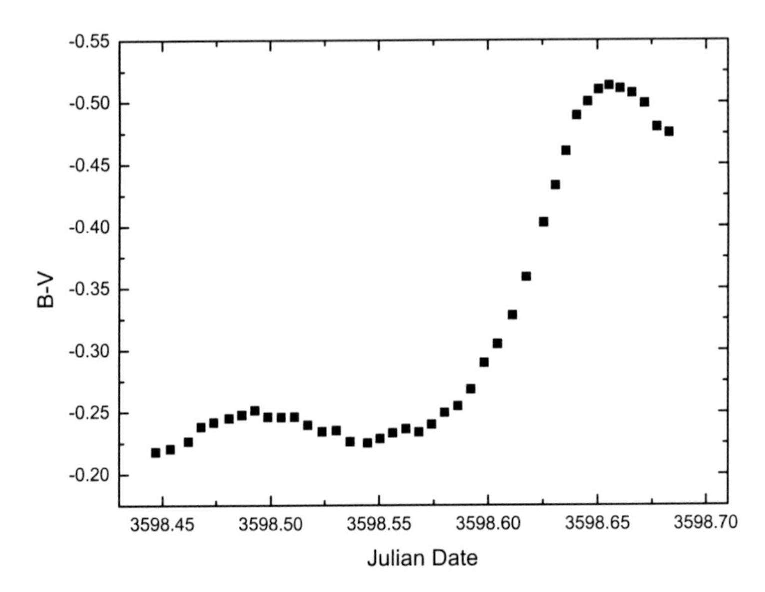


Figure 4: *B-V* diagram of a selected night for SS For. Julian Date is given in the form 2450000+. The bump is clearly visible and therefore stronger in Johnson *B* than in *V*. Scatter is smaller than the symbols.

Regarding the Blazhko phase, however, we clearly see that in SS For the bump is most distinct during Blazhko minimum and almost vanishes around Blazhko maximum. For RR Lyr on the other hand previous authors such as Walraven (1949) and Fringant (1961) reported opposite behavior, finding the bump more conspicious during maximum amplitude than during minimum. We cannot confirm this as our investigation on the bump strength in RR Lyr does not allow us to conclude any noteworthy dependance on the Blazhko phase. Generally, one has to remark that it is very difficult to find a objective measure of the bump strengh as during its progress through the phase diagram it moves up and down the descending branch of the light curve and at the same time changes its width. In our analysis we used two different methods to determine the strength of the bump. First we simply measured the brightness of the bump maximum ignoring the motion along the descending branch. The second, more sophisticated method involved brightness measurements both left and right of the bump from which an average value was calculated. The bump amplitude then was defined as the difference between the bump maximum and this value. Both methods yielded comparable results.

## Conclusions

In in this paper we examined the behaviour of the bump in the light curve over the Blazhko cycle for two stars, SS For and RR Lyr. SS For is characterized by strong variations around minimum light. In both examined stars, the bump appearing just before minimum light moves back and forth in the phase diagram, with a period equal to the Blazhko period of the star. We conclude that there is a direct connection between the Blazhko effect and the motion of the bump. Models for the bump should therefore allow the bump phase to be variable. It will be necessary to investigate a larger sample of Blazhko and non-Blazhko stars to find out whether a variable bump is a common phenomenon.

**Acknowledgments.** The authors thank Michel Breger and the members of the TOPS team (Theory and Observation of Pulsating Stars) for fruitful discussions on the topic. Part of this investigation has been supported by the Austrian Fonds zur Förderung der wissenschaftlichen Forschung, project number P17097-N02.

## References

Blazhko, S. 1907, Astr. Nachr., 175, 325
Carney, B.W., Storm, J., Trammell, S.R., & Jones, R.V. 1992, PASP, 104, 44-56
Fringant, A. 1961, J.Obs., 44,165
Gillet, D., & Crowe, R.A. 1988, A&A, 199, 242-254
Hill, S.J. 1972, Astrophys. J., 178, 793
Kolenberg, K., Smith, H.A., Gazeas, K.D., et al. 2006, A&A, in press
Lenz, P., & Breger, M. 2005, CoAst, 146, 53
Lub, J. 1977, Astron. Astrophys. Suppl., 29, 345
Pojmanski, G. 2002, Acta Astronomica, 50, 177
Walraven, T. 1949, Bull. Astr.Inst., 11, 17

*Comm. in Asteroseismology*
*Vol. 148, 2006*

# HD 114839 - An Am star showing both $\delta$ Scuti and $\gamma$ Dor pulsations discovered through MOST photometry

H. King[1], J.M. Matthews[1], J.F. Rowe[1],
C.Cameron[1], R.Kuschnig[1], D.B. Guenther[2], A.F.J. Moffat[3], S.M. Rucinski[4],
D. Sasselov[5], G.A.H. Walker[1], W.W. Weiss[6]

[1] Department of Physics and Astronomy, University of British Columbia
6224 Agricultural Road, Vancouver, British Columbia, Canada, V6T 1Z1
[2] Department of Astronomy and Physics, St. Mary's University
Halifax, NS B3H 3C3, Canada
[3] Département de physique, Université de Montréal
C.P. 6128, Succ. Centre-Ville, Montréal, QC H3C 3J7, Canada
[4] David Dunlap Observatory, University of Toronto
P.O. Box 360, Richmond Hill, ON L4C 4Y6, Canada
[5] Harvard-Smithsonian Center for Astrophysics
60 Garden Street, Cambridge, MA 02138, USA
[6] Institut für Astronomie, Universität Wien
Türkenschanzstrasse 17, A-1180 Wien, Austria

## Abstract

Using MOST[1] (Microvariability and Oscillations of STars) satellite guide star photometry, we have discovered a metallic A star showing hybrid p- and g-mode pulsations. HD 114839 was observed nearly continuously for 10 days in March, 2005. We identify frequencies in three groups: the first centered near 2 cycles/day, in the $\gamma$ Dor pulsation range, and two others near 8 and 20, both in the $\delta$ Scuti range. This is only the fourth known such hybrid pulsator, including another MOST discovery (Rowe et al. 2006).

[1]MOST is a Canadian Space Agency mission, operated jointly by Dynacon, Inc., and the Universities of Toronto and British Columbia, with assistance from the University of Vienna.

## Introduction

Delta Scuti variables are A-F type stars exhibiting both radial and non-radial p-mode pulsations with periods ranging from about 1 to 5 hours. Gamma Doradus variables are non-radial g-mode pulsators with periods ranging from about 7 hours to 3 days (Kaye et al. 1999), which overlap the red edge of the $\delta$ Scuti instability strip (Handler 1999).

This region of the HR Diagram was searched by Handler & Shobbrook (2002) for hybrid pulsators exhibiting both p- and g-modes. Their search entailed about 270 hr of photometry of a sample of 26 known and candidate $\gamma$ Dor stars, to look for $\delta$ Scuti-type oscillations. They discovered one hybrid pulsator, HD 209295, a binary system for which the authors argue that excitation of the $\gamma$ Dor g-modes is likely due to tidal interaction. The first known example of a single star showing both $\delta$ Scuti and $\gamma$ Dor pulsations is the metallic A (Am) star HD 8801 (Henry & Fekel 2005). The authors identified three distinct pairs of frequencies consistent with g- and p-modes.

It was once thought that all Am stars (which do not have strong global magnetic fields) had binary companions as the measured values of vsini were systematically lower than those for non-peculiar stars in the same part of the HR diagram. The proposed mechanism was tidal breaking in short-period ($P \leq 100$ days) binary systems. Even in the absence of a stabilising magnetic field, the slow rotation might reduce meridional circulation and other atmospheric turbulence enough to allow chemical diffusion to produce the observed abundance anomalies (Abt & Levy 1985). In Abt & Levy's study of a sample of 55 Am stars, they found that 75% showed evidence of spectroscopic binarity. However, the authors estimated that only an additional 8% of their sample were spectroscopic binaries that were missed due to low orbital inclinations, suggesting that in fact not all Am stars are in binary systems. Another possibility was that slow rotation in single and long-period binary Am stars could be explained by evolutionary effects, but more recent studies (see Henry & Fekel 2005) suggest that there must be additional factors that have yet to be determined.

Asteroseismology of Am stars could help determine or constrain the physical parameters and evolutionary states of these peculiar stars. Only a small number of pulsation frequencies has been detected in HD 8801 (Henry & Fekel 2005), limiting its potential for seismic modelling. To explore the possibility of such modelling, other pulsating Am stars with richer eigenspectra must be found.

We report here on just such a discovery, made with the MOST (Microvariability & Oscillations of STars) space mission (Matthews et al. 2004, Walker et al. 2003). The MOST satellite (launched on 30 June 2003) houses a 15-cm telescope feeding a CCD photometer through a custom broadband optical filter. Its primary mission was to obtain very high-precision photometry of bright stars ($V \leq 6$) to detect pulsations with amplitudes as low as a few $\mu$mag, with sampling rates of better than one exposure per minute and nearly continuous coverage for weeks at a time. MOST's capabilities were improved and extended after launch to enable astronomical photometry of the fainter ($11 \leq V \leq 7$) guide stars in the target fields, used to control spacecraft pointing.

During March 2005, HD 114839 was one of the guide stars for observations of the MOST Primary Science target $\beta$ Comae, and low-amplitude oscillations were evident even in the raw photometry. HD 114839 had not attracted much attention prior to these MOST observations. The only accurate stellar parameters for this star available in the literature are from the Hipparcos catalogue: $V = 8.46 \pm 0.01$, $B - V = 0.31 \pm 0.01$, spectral type = Am, parallax $\pi = 5.04 \pm 1.04$ mas, and $M_V = 2.06 \pm 0.45$.

## Photometry and frequency analysis

HD 114839 was monitored nearly continuously for about 10 days during 22 - 31 March 2005, with only one short gap of about 5 hours during the entire run. The exposure time was 25 sec, sampled every 30 sec. Our final data set contains 27,540 measurements with a duty cycle of 93.5%.

As a guide star, HD 114839 was centred on a $45 \times 45$-pixel subraster of the MOST Startracker CCD. The guide star photometry is almost entirely preprocessed on board the spacecraft (see Walker et al. 2005 for additional details). The mean of the top and bottom rows of the subraster is calculated to provide a threshold value of that mean plus 30 ADU (Analogue-to-Digital Units). The intensities of all pixels in the subraster above that threshold are summed on board, and this value is downloaded to Earth as a flux value for the star, with crude sky subtraction.

The sky background is typically modulated by the 101.4-min orbital period of the MOST satellite, due to stray Earthshine (see Reegen et al. 2005 and Rowe et al. 2006). Since no true sky background measurements are available for MOST guide star photometry, we subtract from the data a running average of the background phased with the MOST orbital period (see Rucinski et al. 2004). This automatically suppresses the MOST orbital frequency and its harmonics in the Fourier spectrum of the data.

The frequency analysis for this star was performed with the program CAPER (Walker et al. 2005) developed by one of the authors (CC). CAPER uses a Discrete Fourier Transform (DFT) to identify frequencies and amplitudes, and then obtains a solution via a simultaneous non-linear least-squares (Leveberg-Marqardt) fit (Press et al. 1986). We refer the reader to Saio et al. (2006) and Cameron et al. (2006) for more details of CAPER.

The Fourier amplitude spectrum of the HD 114839 photometry is plotted in Figure 1, along with the spectral window (which is very clean due to the high duty cycle of the 10 days of MOST data). The initial frequency analysis identified 22 significant frequencies, of which 7 can be attributed to instrumental or orbital artifacts like the stray light modulation mentioned above. Because the photometry is nondifferential, we conservatively reject all power in the DFT below 0.5 cycles/day (c/d). Comparison of the Fourier spectrum of the HD 114839 data to the DFTs of three other guide stars (fainter by about 1 to 1.5 mag) in the same field shows clearly that the peaks we identify at frequencies above about 1 c/d are unique and intrinsic to HD 114839.

The frequencies and amplitudes are listed in Table 1.

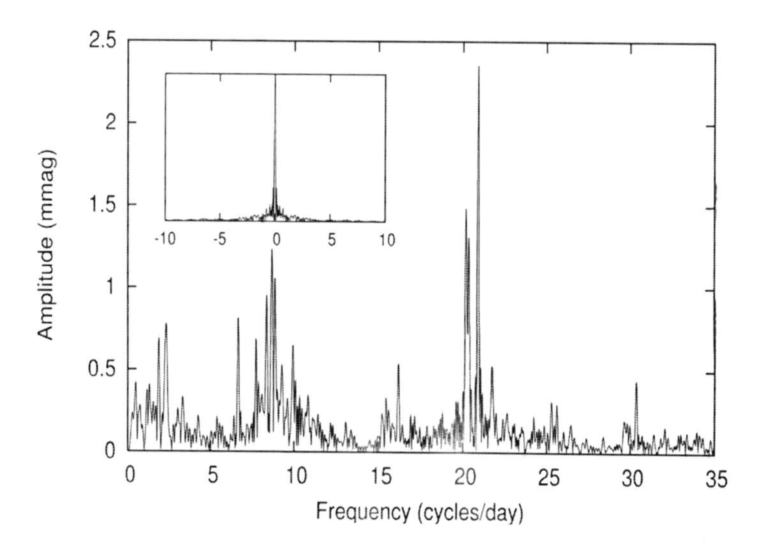


Figure 1: Fourier Spectrum of HD 114839, after removal of known instrumental artifacts. Inset is the window function for the highest peak in the spectrum.

## A new hybrid pulsator

We identify 15 frequencies in HD 114839 with a signal-to-noise (S/N) level in amplitude above 3.8. Four of these frequencies are between 1 and 2.5 c/d, consistent with $\gamma$ Dor g-mode pulsations, while the remaining frequencies are $\delta$ Scuti-type p-modes between about 6.5 and 22 c/d. The p-mode frequencies are grouped into two ranges, 7 frequencies around 8 c/d and 4 around 21 c/d. We note that the frequencies found by Henry & Fekel (2005) in HD 8801

Table 1: Identified frequencies for the star HD 114839

| Frequency(day$^{-1}$) | Amplitude(mmag) | S/N |
|---|---|---|
| 1.3412 | 0.473±0.004 | 3.87 |
| 1.8905 | 0.637±0.004 | 4.25 |
| 2.2609 | 0.609±0.004 | 3.95 |
| 2.3376 | 0.771±0.004 | 4.75 |
| 6.6678 | 0.794±0.004 | 4.87 |
| 7.7152 | 0.624±0.004 | 5.22 |
| 8.3411 | 0.910±0.004 | 6.94 |
| 8.6477 | 1.210±0.004 | 8.84 |
| 8.8649 | 1.000±0.004 | 8.00 |
| 9.2864 | 0.528±0.004 | 4.32 |
| 9.9378 | 0.593±0.004 | 4.97 |
| 20.1822 | 1.520±0.004 | 9.58 |
| 20.3483 | 1.450±0.004 | 10.02 |
| 20.8976 | 2.260±0.004 | 15.23 |
| 21.7534 | 0.461±0.004 | 4.72 |

– another Am star – were clustered in a similar fashion, near 3, 8 and 20 c/d. We cannot establish without doubt that HD 114839 is a single star, lacking spectroscopic confirmation, but the possible parallels with HD 8801 are intriguing. In any event, this discovery and that by Rowe et al. (2006) reported in this issue together double the number of known hybrid pulsators, and possibly triple the number of single stars which exhibit both $\delta$ Scuti and $\gamma$ Doradus modes simultaneously.

**Acknowledgments.** JMM, DBG, AFJM, SR, and GAHW are supported by funding from the Natural Sciences and Engineering Research Council (NSERC) Canada. RK is funded by the Canadian Space Agency. WWW is supported by the Austrian Science Promotion Agency (FFG - MOST) and the Austrian Science Funds (FWF - P17580).

## References

Abt, H.A, & Levy, S.G. 1985, ApJS, 59, 229
Cameron, C., Matthews, J.M., Rowe, J.F., et al. 2006, CoAst, 148, 58
ESA 1997, The HIPPARCOS and TYCHO catalogues, Astrometric and photometric star catalogues derived from the ESA HIPPARCOS Space Astrometry Mission (ESA Publikations Division, Noordwijk, Netherlands), ESA SP Ser., 1200

Handler, G. 1999, MNRAS, 309, L19
Handler, G., & Shobbrook, R.R. 2002, MNRAS, 333, 251
Henry, G. & W. Fekel, F. C. 2005, AJ, 129, 2026
Kaye, A.B., et al. 1999, PASP, 116, 558
Matthews, J.M., et al. 2004, Nature, 430, 51-53
Press, W.H., et al. 1986, Numerical Recipes in Fortran 77 (Cambridge University Press), 678
Reegen, P., et al. 2005, MNRAS, 367, 1417
Rowe, J.F., Matthews, J.M., Cameron, C., et al. 2006, CoAST, CoAst, 148, 34
Rucinski, S.M., et al. 2004, PASP, 116, 1093
Saio, H., et al. 2006, astro-ph/0606712
Walker, G. A., et al. 2003, PASP, 115, 1023
Walker, G. A., et al. 2005, ApJ, 635, L77

*Comm. in Asteroseismology*
*Vol. 148, 2006*

# Discovery of hybrid $\gamma$ Dor and $\delta$ Sct pulsations in BD+18 4914 through MOST spacebased photometry

J.F. Rowe[1], J.M. Matthews[1], C. Cameron[1], D.A. Bohlender[7], H. King[1], R. Kuschnig[1], D.B. Guenther[2], A.F.J. Moffat[3], S.M. Rucinski[4], D. Sasselov[5], G.A.H. Walker[1], W.W. Weiss[6]

[1] Department of Physics and Astronomy, University of British Columbia
6224 Agricultural Road, Vancouver BC V6T 1Z1
[2] Department of Astronomy and Physics, St. Mary's University
Halifax, NS B3H 3C3, Canada
[3] Département de physique, Université de Montréal
C.P. 6128, Succ. Centre-Ville, Montréal, QC H3C 3J7, Canada
[4] David Dunlap Observatory, University of Toronto
P.O. Box 360, Richmond Hill, ON L4C 4Y6, Canada
[5] Harvard-Smithsonian Center for Astrophysics
60 Garden Street, Cambridge, MA 02138, USA
[6] Institut für Astronomie, Universität Wien
Türkenschanzstrasse 17, A–1180 Wien, Austria
[7] National Research Council of Canada, Herzberg Institute of Astrophysics,
5071 West Saanich Road, Victoria, BC V9E 2E7, Canada

## Abstract

We present a total of 57 days of contiguous, high-cadence photometry (14 days in 2004 and 43 in 2005) of the star BD+18 4914 obtained with the MOST[1] satellite. We detect 16 frequencies down to a signal-to-noise of 3.6 (amplitude $\sim$ 0.5 mmag). Six of these are less than 3 cycles/day, and the other ten are between 7 and 16 cycles/day. We intrepret the low frequencies as g-mode $\gamma$ Doradus-type pulsations and the others as $\delta$ Scuti-type p-modes, making BD+18 4914 one of the few known hybrid pulsators of its class. If the g-mode pulsations are high-overtone non-radial modes with identical low degree $\ell$, we can assign a unique mode classification of $n$=\{12, 20, 21, 22, 31, 38\} based on the frequency ratio method.

[1] MOST is a Canadian Space Agency mission, operated jointly by Dynacon, Inc., and the Universities of Toronto and British Columbia, with assistance from the University of Vienna.

## Introduction

$\gamma$ Doradus stars pulsate with typical periods of about 0.8 days (Kaye et al. 1999), consistent with high-overtone nonradial g-modes. They represent one of the newest classes of pulsating variable stars, and about half of the currently known $\gamma$ Doradus stars lie within the $\delta$ Scuti instability strip (Handler 2005). The $\delta$ Scuti variables exhibit p-modes of low radial order, seen only in low degree photometrically, with typical periods of a few hours. Handler & Shobbrook (2002) have shown that the pulsation characteristics of the two classes can be clearly separated by their values of the pulsation constant Q.

The overlap in physical properties of $\gamma$ Doradus and $\delta$ Scuti stars suggested the possibility that hybrid pulsators may exist. The astroseismic implications are exciting since the g-modes would probe the deep interior of the star and the p-modes, its envelope. This has lead to photometric monitoring of $\gamma$ Doradus and $\delta$ Scuti stars to search for hybrid behaviour. The first such hybrid to be discovered was in the binary system HD 209295 by Handler et al. (2002), from careful monitoring of 26 $\gamma$ Doradus variables (Handler & Shobbrook 2002) but the $\gamma$ Doradus pulsations in the primary component are likely caused by tidal interactions with the secondary. The first convincing case of a single hybrid star, the Am star HD 8801, was discovered by Henry & Fekel (2005) from monitoring 39 stars from a volume-limited sample of 114 $\gamma$ Doradus candidates. HD 8801 shows frequencies clustered around 3, 8 and 20 cycles/day (c/d). There were only two frequencies in the $\gamma$ Doradus range and 4 frequencies classified as $\delta$ Scuti in nature. The low number of frequencies makes this star a challenging subject for asteroseismic modeling, but just the existence of a hybrid single star, and the chemical peculiarity of HD 8801, point to new and interesting astrophysics.

We present photometry of the star BD+18 4914 ($\alpha = 22^h02^m38^s$, $\delta = +18^o54'03''$ [J2000], V=10.6, B=11.1)[2] by the MOST (Microvariability & Oscillations of STars) satellite in which we detect frequencies consistent with hybrid pulsations.

## Photometry

MOST (Walker, Matthews et al. 2003) is a microsatellite housing a 15-cm telescope feeding a CCD photometer through a custom broadband optical filter. Launched in June 2003 into an 820-km circular Sun-synchronous polar orbit (period = 101.413 min), MOST can monitor stars in its Continuous Viewing Zone (CVZ) for up to 8 weeks without interruption. It collects photometry in

[2]This research has made use of the SIMBAD database, operated at CDS, Strasbourg, France.

three ways: (1) Fabry Imaging, projecting an extended image of the telescope pupil illuminated by a bright target (see Matthews et al. 2004; Reegen et al. 2005); (2) Guide Star photometry, based on onboard processing of faint stars used for telescope pointing (see Walker et al. 2005); and (3) Direct Imaging, where defocused star images (FWHM $\sim$ 2.5 pixels) are projected onto an open area of the Science CCD. This last technique was used to obtain the BD+18 4914 photometry, and details about the Direct Imaging process and reduction are provided by Rowe et al. (2006b); hereafter, RMSK.

The observations of BD+18 4914 were carried out during 14-30 August 2004 and 1 Aug - 15 Sept 2005, for a total of 57 days, during Direct Imaging photometry of the transiting exoplanet system HD 209458 (RMSK). Panel D of Figure 1 shows the 2005 observations of BD+18 4914 using 40-min bins to clearly demonstrate the low-frequency pulsations with periods around 1 day.

The exposure time was 1.5 sec, sampled once every 10 seconds. The 2005 data have a raw duty cycle of 97.3%, with about 360 000 measurments in 43 days. After rejection of points with extreme cosmic ray activity and other obvious outliers, the net duty cycle is 75%. The 2004 data have a net duty cycle of 70.4%. The combined data set have a total of 443 154 measurements in 57 days. There are gaps about 30 minutes long during each 101.4 minute orbit of the satellite. The data do not suffer from the cycle/day aliases common to groundbased photometry. They have only minor alias sidelobes at 14.2 c/d (see the spectral window function in Figure 1) which do not lead to any ambiguities in the frequency identifications.

The photometric reduction scheme is described in RMSK for the 2004 data set. A paper on the data of 2005 is in preparation. After dark and flatfield corrections, photometry is obtained by a combination of aperture measurments for the core of the stellar point spread function (PSF) and a fit to the Moffat-profile (Moffat 1969) model for the wings of the PSF.

The observations are made through a single broadband filter (350 - 750 nm) made specifically for the MOST mission, which has about $3\times$ the throughput of a Johnson V bandpass but is not tied to any standard photometric system.

## Frequency Analysis

A preliminary analysis of the 2004 observations of BD+18 4914 was presented in Rowe et al. (2006a) that showed the dual nature of its pulsations. Here we take a more methodological approach to the frequency analysis.

We begin by computing the discrete Fourier transform (DFT) of the 2005 time series. The amplitude spectrum is shown in panel A of Figure 1. The data

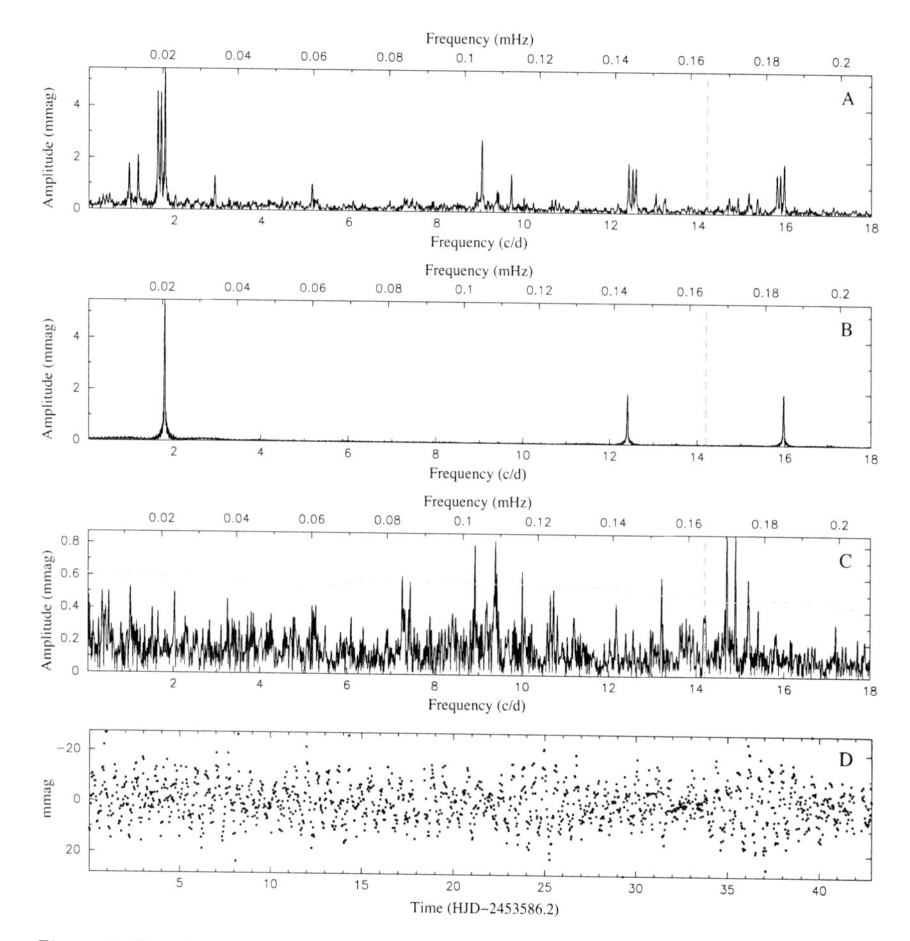


Figure 1: Panel A shows the amplitude spectrum of the 2005 data set. Panel B shows the corresponding window function for the highest peak in Panel A. Panel C shows the amplutude spectrum after prewhitening of the 8 strongest frequencies. The detection limit of 3.6 times the noise is also shown. (The MOST orbital frequency is marked as a dashed vertical line.) Panel D presents the photometric data in 40-min bins to highlight the low-frequency (2 c/d) oscillations.

are then fitted using an equation of the form

$$mag = A_0 + \sum_{j=1,n} A_j \cos(2\pi f_j t + \phi_j), \tag{1}$$

Where $A_0$ is a linear offset and $f_j$, $A_j$ and $\phi_j$ are the frequency, amplitude and phase for each successive peak found in the amplitude spectrum using the

Leveberg-Marqardt approach (Press et al. 1992). After each fit, the DFT of the residuals is recalculated and the next largest amplitude is chosen, but only if the Signal-to-Noise (S/N) is greater than 3.6. The S/N is defined as the amplitude of the peak in the spectrum divided by the mean of nearby frequencies. We use a window about 3 c/d wide in frequency space, centred on the highest peak, to calculate the mean which we use as an estimate of the local noise floor.

We detect 16 frequencies in the 2005 photometry with a S/N greater than 3.6. Our results and best fit parameters are listed in Table 1.

If the same procedure is repeated for the 2004 photometry, then the relatively short duration (14 days) of the data set causes degeneracies in the nonlinear solution because of poor frequency resolution. Specifically, the solutions for frequencies $j = \{2, 3\}$ in Table 1 converge to identical values with phases offset by $\pi$ radians. The amplitudes in turn grow unreasonably large. The problem is illustrated in Figure 2. The two top panels of Figure 2 compare the amplitude spectra of the 2005 and 2004 data sets in the frequency range 0.1 - 3 c/d. The 2005 data set has a higher frequency resolution due to its longer duration. The 2004 amplitude spectrum shows the same three peaks as seen in 2005, but the amplitudes of the peaks appear to be larger in amplitude. This is due to the degeneracy of the solution at these low frequencies in the shorter time series. The two bottom panels of Figure 2 compare the DFT at a higher frequency range, 8 - 11 c/d. The frequency and amplitude changes seen in the low frequency range are no longer apparent.

To avoid the degeneracy problem, the frequency solution from the 2005 dataset is used as the initial solution for the non-linear routine applied to the 2004 data, with the frequencies held as fixed parameters to derive the amplitudes to be determined. If one examines the DFT of the residuals from the best fit there are no significant peaks remaining, thus the 2005 frequencies solution is valid for the 2004 data set. The best fit parameters are presented in column 5 of Table 1. This does not imply that the frequencies are constant from 2004 to 2005, only that the 2004 data set is too short to give meaningful results for a a change in frequeny in the low frequency regime.

To estimate the errors in our fitted parameters, we perform a "bootstrap" analysis. This method involves redetermining the fitted parameters with randomly generated data sets. The new datasets are created by randomly selecting data from the original time series with replacement. In other words, any individual data point can be chosen more than once but the total number of selected points is always the same as the original data set. The bootstrap method is effective since it preserves the same noise profile in each random set as exists in the original data and given enough iterations will produce error distributions for each fitted variable. For further demonstrations and discussion of the bootstrap method, we refer the reader to Cameron et al. (2006) and Saio et al. (2006).

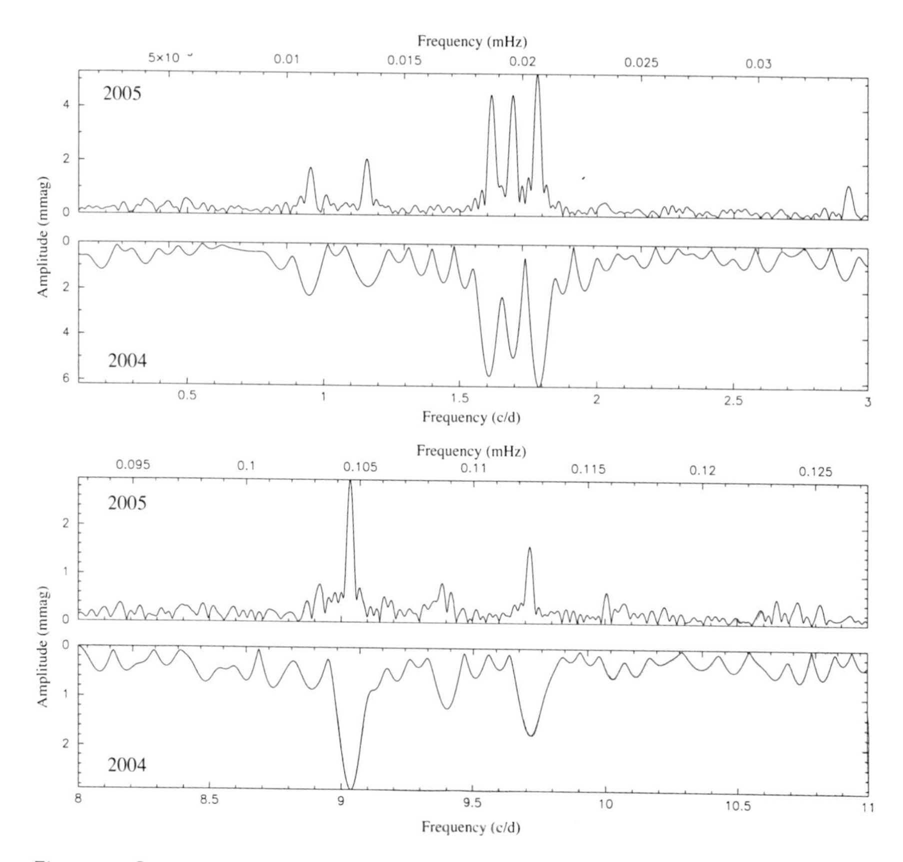


Figure 2: Comparisons of the 2004 and 2005 Fourier amplitude spectra. The two top panels compare the two data sets in the frequency range 0 - 3 c/d. The two bottom panels cover the range 8 - 11 c/d.

We generated 22083 and 18595 bootstrap iterations for the 2004 and 2005 data sets, respectively. The 1-$\sigma$ error distributions using a Gaussian model are given in Table 1 for both data sets.

## A hybrid pulsator

The frequencies found in BD+18 4914 cluster in the two ranges typical of $\gamma$ Dor and $\delta$ Sct oscillation modes, making this a clear candidate for a hybrid pulsator.

We can quantify this assessment using the criterion established by Handler & Shobbrook (2002) that the pulsation constant Q distinguishes the $g-$ and $p-$

| $j$ | $f_j$ (c/d) | $A_j$ (mmag) | $\phi_j$ (rad) | $A_j^{2004}$ (mmag) | S/N | Q (days) |
|---|---|---|---|---|---|---|
| | $\sigma_{f_j}$ | $\sigma_{A_j}$ | $\sigma_{\phi_j}$ | $\sigma_{A_j^{2004}}$ | | |
| 6 | 0.9496 | 1.97 | 6.08 | 1.65 | 10.0 | 0.34 |
| | 0.0004 | 0.06 | 0.07 | 0.10 | | |
| 5 | 1.1586 | 2.07 | 3.60 | 1.75 | 9.9 | 0.28 |
| | 0.0004 | 0.06 | 0.07 | 0.10 | | |
| 2 | 1.6150 | 4.74 | 0.55 | 4.57 | 19.0 | 0.20 |
| | 0.0002 | 0.06 | 0.03 | 0.10 | | |
| 3 | 1.6924 | 4.60 | 2.05 | 4.48 | 20.0 | 0.19 |
| | 0.0002 | 0.06 | 0.03 | 0.10 | | |
| 1 | 1.7829 | 5.25 | 5.99 | 4.98 | 19.4 | 0.18 |
| | 0.0002 | 0.06 | 0.03 | 0.10 | | |
| 8 | 2.9286 | 1.22 | 3.08 | 0.87 | 7.0 | 0.11 |
| | 0.0007 | 0.06 | 0.12 | 0.10 | | |
| 14 | 7.2530 | 0.65 | 0.22 | 1.04 | 3.9 | 0.04 |
| | 0.0012 | 0.06 | 0.21 | 0.10 | | |
| 15 | 7.4354 | 0.53 | 0.03 | 0.42 | 3.8 | 0.04 |
| | 0.0015 | 0.06 | 0.27 | 0.10 | | |
| 11 | 8.9122 | 0.84 | 4.66 | 0.94 | 4.9 | 0.04 |
| | 0.0010 | 0.06 | 0.17 | 0.10 | | |
| 4 | 9.0348 | 3.09 | 0.43 | 3.08 | 16.3 | 0.04 |
| | 0.0003 | 0.06 | 0.05 | 0.10 | | |
| 12 | 9.3847 | 0.84 | 0.90 | 1.10 | 5.1 | 0.03 |
| | 0.0010 | 0.06 | 0.17 | 0.10 | | |
| 7 | 9.7156 | 1.56 | 3.95 | 1.85 | 9.4 | 0.03 |
| | 0.0005 | 0.06 | 0.09 | 0.10 | | |
| 13 | 10.0043 | 0.60 | 3.09 | 0.50 | 3.9 | 0.03 |
| | 0.0013 | 0.06 | 0.23 | 0.10 | | |
| 10 | 14.6977 | 0.91 | 3.07 | 0.64 | 7.0 | 0.02 |
| | 0.0010 | 0.06 | 0.16 | 0.10 | | |
| 9 | 14.8967 | 0.87 | 3.09 | 0.85 | 6.9 | 0.02 |
| | 0.0010 | 0.06 | 0.16 | 0.10 | | |
| 16 | 15.4106 | 0.46 | 3.76 | 0.24 | 3.8 | 0.02 |
| | 0.0018 | 0.06 | 0.31 | 0.09 | | |

Table 1: Observed frequencies and parameters for BD+18 4914. The epoch is HJD=2453586.20349121.

modes in this type of star. It was shown that although the pulsation periods of $\delta$ Scuti and $\gamma$ Doradus overlap there is a clear separation when Q is considered (see Figure 9 of Handler & Shobbrook (2002)). To calculate Q, we require basic properties of the star: log $g$, $M_{bol}$ and $T_{eff}$. We obtained a 10Å/mm spectrum with the 1.8m Plasket telescope at the Dominion Astrophysical Observatory[3] covering a range of 6473-6716Å. Our initial analysis gives values of $T_{eff}$ = 7250 K, log $g = 3.7$ cgs and $M_{bol} = 2.5$. Using Equation 1 from Handler & Shobbrook (2002), we compute values of Q for each frequency, and they are presented in Table 1.

With analogy to the Am star hybrid pulsator HD 8801 (Henry & Fekel 2005), we assume that all frequencies less than 3.0 c/d are of $\gamma$ Doradus type and the frequencies higher than 6 c/d are of $\delta$ Scuti type. Although we do not have enough information (multibandpass photometry or spectral line profile variation data) to make pulsation mode identifications, we can apply the Frequency Ratio Method (FRM) described by Moya et al. (2005) to the 6 lowest frequencies. This assumes that the observed $\gamma$ Dor pulsations can be described by the asymptotic approximation under the assumption of adiabaticity and spherical symmetry (Tassoul 1980). If the modes all share the same degree $\ell$, then the ratio of the frequencies can be approximated by

$$\frac{\sigma_{\alpha 1}}{\sigma_{\alpha 2}} \approx \frac{n_2 + 1/2}{n_1 + 1/2}, \qquad (2)$$

Under these assumptions, we have searched for 6 overtone $n$ values which satisfy Equation 2, taking an error of $\pm 1.3 \times 10^{-2}$ for calculation of the sets of possible overtones (Suárez et al. 2005). Restricting our search to overtones up to and including $n = 60$, we find only one viable solution: $n = \{12, 20, 21, 22, 31,$ and $38\}$ for the frequencies labeled $j = \{8, 1, 3, 2, 5,$ and $6\}$ in Table 1. If we search up to $n = 100$, then the density of natural number ratios compared to our error bounds allows for 78 more solutions. Regardless, other mode identification methods need to be applied to restrict the possible values of degree $\ell$.

Observations of other $\gamma$ Doradus pulsators have shown that the amplitudes can be variable, such as seems to be the case with 9 Aurigae (see Kaye et al. 1997 and references therein). Using our results for the best-fit parameters of the 2004 and 2005 photometry, we can examine the possibility of amplitude changes over a 1-year interval. In Figure 3 we plot the measured amplitudes from 2004 versus those from 2005 (see Table 1) with $1\sigma$ error bars. No significant amplitude changes have occurred. We define detection of an amplitude change

[3] Based in part on observations obtained at the Dominion Astrophysical Observatory, Herzberg Institute of Astrophysics, National Research Council of Canada

as

$$\frac{\Delta A_j}{\sigma} = \frac{A_j - A_j^{2004}}{\sqrt{\sigma_{A_j}^2 + \sigma_{A_j^{2004}}^2}} \tag{3}$$

where the definitions are same as presented in Table 1. The distribution of $\Delta A_j/\sigma$ appears to be non-Gaussian. We can test this with an student T-test and an F-test. To do so, we generated 10000 sets of 16 random Gaussian deviates and calculated the student T-test probability and F-test probability for the distributions to have similar means to our $\Delta A_j/\sigma$ sample. Our adapted criteria (for the samples to differ) is a probability less than 0.0026 (3 $\sigma$). For the T-test, none of the 10000 sets were rejected; thus the means are statistically similar. For the F-test, 26.7% of cases were rejected (i.e., only 27.6% produced a probability less than 0.0026). At our chosen $3\sigma$ threshold, our sample has a Gaussian distribution thus an amplitude change is not statistically significant. If we apply the tests to our 2004 and 2005 amplitude distributions and ask if the two distributions differ, then the student T-test gives a probability of 90.9% and the F-test, a probability of 88.3% that the two have similar means and variances, respectively.

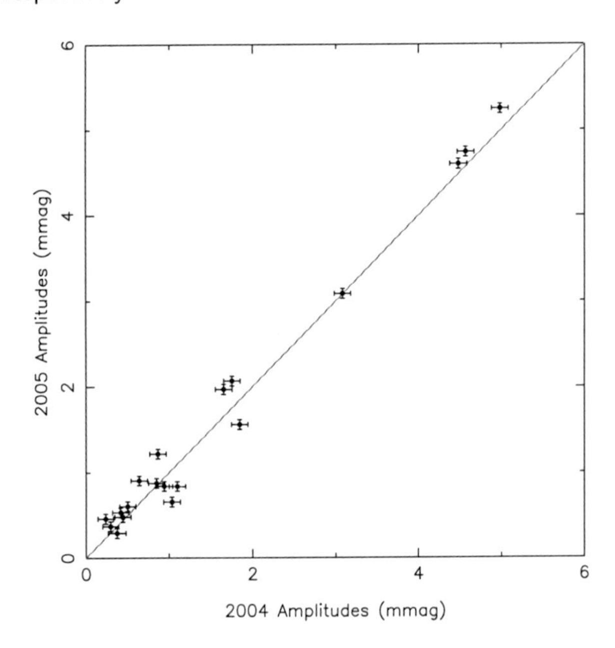


Figure 3: The measured amplitudes for the 2004 and 2005 observation campaigns plotted against each other, with $1\sigma$ error bars.

## Conclusions

We have presented 14 and 43 days of nearly continuous photometry obtained by the MOST satellite. From this photometry, we detect 16 frequencies whose amplitudes have $S/N > 3.6$, clustered in two ranges consistent with $\gamma$ Doradus-type and $\delta$ Scuti-type pulsations. With 6 frequencies in the $\gamma$ Doradus range, application of the FRM method assuming a common degree $\ell$ yields a unique set of radial orders of the pulsations. Comparison of the 2004 and 2005 data sets show no statistically significant changes in the pulsation amplitudes. However, this star is scheduled to be observed for a third time by MOST in the fall of 2006 to gain insight into the stability of the observed frequencies and remove any degeneracies from our fits. Groundbased spectroscopy and multicolour photometry will be necessary to obtain independent mode identifications to confirm whether the FRM assumptions we have made are valid and to take advantage of the potential of BD+18 4914 for asteroseismology.

**Acknowledgments.** The contributions of JMM, DBG, AFJM, SR, and GAHW are supported by funding from the Natural Sciences and Engineering Research Council (NSERC) Canada. RK is funded by the Canadian Space Agency. WWW received financial support from the Austrian Science Promotion Agency (FFG - MOST) and the Austrian Science Funds (FWF - P17580).

## References

Cameron, C., et al. 2006, CoAst, 148, 34
Handler, G. & Shobbrook, R.R. 2002, MNRAS, 333, 251
Handler, G., et al. 2002, MNRAS, 333, 262
Handler, G. 2005, JApA, 26, 241
Henry, G.W. & Fekel, F.C. 2005, AJ, 129, 2026
Kaye, A. B., et al. 1997, DSSN, 11, 32
Kaye, A. B., et al. 1999, PASP, 116, 558
Matthews et al. 2004, Nature, 430, 51
Moffat, A.F.J. 1969, A&A , 3, 455
Moya, A. et al. 2005, A&A , 432, 189
Press, W.H. et al. 1992, Numerical Recipes in FORTRAN 77 (2nd ed, Cambridge: Cambridge Univ. Press)
Reegen, P., et al. 2005, MNRAS, 367, 1417
Rowe, J.F., et al. 2006a, Mm. S.A.It., 77, 282
Rowe, J.F., et al. 2006b, ApJ , 646, 1241 (RMSK)
Saio, H., et al. 2006, astro-ph/0606712
Suárez et al. 2005, A&A , 443, 271
Tassoul, M. 1980, ApJ, 43, 469
Walker, G.A.H., Matthews, J.M., et al. 2003, PASP, 115, 1023
Walker, G.A.H., et al. 2005, ApJ, 635, 77

*Comm. in Asteroseismology*
*Vol. 148, 2006*

# 1 Mon: The 'lost' 1979 - 1985 data

M. Breger, Yu. T. Fedotov [1]

Institut für Astronomie, Türkenschanzstrasse 17, 1180 Vienna, Austria

## Abstract

Extensive photoelectric measurements of the $\delta$ Scuti variable 1 Mon were obtained from 1979 to 1985 at Odessa Astronomical Observatory. These data are presented and analyzed. The star shows an almost equidistant frequency triplet with a number of combinations of these three frequencies. The frequencies and amplitudes are in agreement with those found in previous years.

## Introduction

The $\delta$ Scuti variable 1 Mon is of special interest because of its low projected rotational speed of 19 km/s and peak-to-peak amplitude of just under 0.3 mag. This puts the star in the intermediate region between the slowly rotating HADS (high-amplitude Delta Scuti stars) and the ocean of more rapidly rotating Delta Scuti stars with small amplitudes and mostly nonradial pulsation. The examination of these intermediate stars (e.g., 1 Mon and 44 Tau) is one of the present priorities of the Delta Scuti Network.

1 Mon is special for another reason: the star shows an almost equidistantly spaced frequency triplet. However, this triplet is composed of modes with different quantum numbers, $\ell$ (Balona & Stobie 1980, Balona et al. 2001). Consequently, rotational splitting of a nonradial mode as a cause for equidistancy seems to be ruled out.

The variability of 1 Mon was discovered by Cousins (1963). Several subsequent short studies confirmed the variability using a single frequency with variable amplitude. Millis (1973) showed the existence of two frequencies of pulsation. It was the extensive study by Shobbrook & Stobie (1974) which revealed the existence of three frequencies as well as interactions between them.

[1] formerly at Astronomical Observatory, Park Shevchenko. Odessa 270014, Ukraine

To examine the nature of the frequency triplet, lengthy studies are needed: for the 1970/71 and 1971/72 observing seasons (Millis data listed in his paper), for 1972/73 (Shobbrook & Stobie data available from the IAU Archives of Unpublished Observations of Variable Stars), for 1976/77 and 1977/78 (Balona & Stobie 1980, data on microfiche in the journal,[2] while the less extensive photometric data for the year 2000 are not publicly available.

During the 1978/79, 1982/83 and 1984/85 observing seasons, additional measurements of 1 Mon were obtained at the Astronomical Observatory of the Odessa State University. These were announced with very few details (Romanov & Fedotov 1979, Fedotov & Galdyr 1991). It is the purpose of the present paper to provide a brief multiperiod analysis and to make these important data available.

The observations were obtained with the photon-counting photometer and the $V$ filter of the 20-cm refractor at the observational site "Mayaki" of the Astronomical Observatory of the Odessa State University. HR 2039 and HR 2198 were used as comparison stars. All measurements were transformed to the standard $UBV$ system using annual transformation coefficients. Most measurements were carried out during the 1984/85 season, while some are also available from 1979 and 1982/83.

The analysis of the Odessa measurements showed that a few measurements needed to be rejected since they appear to be of very low precision or erroneous. This includes the complete night of 244 5307, for which a very large timing error of the two-hour run is suspected. Altogether, 799 measurements were used.

## Analysis and pulsation behavior of 1 Mon

The photometric data presented here were analyzed for periodicities with the PERIOD04 package (Lenz & Breger 2005), which uses Fourier as well as least-squares algorithms. Data from other longer studies mentioned previously were included to provide a comparison. The previously known three main frequencies as well as a number of combination frequencies were detected. The results for the Odessa data are shown in Table 1. Note that due to zero-point uncertainties (see below) the amplitude for the mode at 0.1291 cd$^{-1}$ needs to be treated with caution and has been put in brackets.The 13-frequency solution fits the observations with a standard deviation of $\pm$ 0.024 mag. However, these deviations are not caused by poor frequency solutions, but mainly by systematic zero-point offsets between measurements taken in different nights. Consequently, we have made an additional solution allowing for nightly zero-point corrections. Such a

[2]Since microfiche data may be difficult to read without special equipment, an electronic version of the data can be obtained from the first author of the present paper.

Table 1: Frequencies and amplitudes of 1 Mon from the Odessa data.

| Frequency $cd^{-1}$ | Name | $V$ amplitude (mag) Zero-point adjustments none | applied |
|---|---|---|---|
| Main modes | | | |
| 7.346153 | $f_1$ | 0.101 | 0.099 |
| 7.475269 | $f_2$ | 0.065 | 0.065 |
| 7.217116 | $f_3$ | 0.020 | 0.018 |
| Other frequencies | | | |
| 6.717240 | $f_4$ | 0.002 | 0.004 |
| 14.821422 | $f_1 + f_2$ | 0.017 | 0.018 |
| 14.692306 | $2f_1$ | 0.013 | 0.012 |
| 14.950538 | $2f_2$ | 0.004 | 0.006 |
| 14.563268 | $f_1 + f_3$ | 0.007 | 0.007 |
| 22.167575 | $2f_1 + f_2$ | 0.004 | 0.003 |
| 22.296691 | $2f_2 + f_1$ | 0.003 | 0.003 |
| 7.604385 | $2f_2$ - $f_1$ | 0.006 | 0.005 |
| 22.038459 | $3f_1$ | 0.001 | 0.002 |
| 0.129116 | $f_2$ - $f_1$ | (0.010) | - |
| Residuals of fit | | ± 0.024 | ± 0.014 |

procedure makes the search for low frequencies impossible and we have therefore omitted the $f_1$-$f_2$ peak at 0.1291 $cd^{-1}$ . Nevertheless, the residuals are now highly improved to ± 0.014 mag.

We detect no amplitude variability between the three observing seasons. Since 75% of the available measurements belong to a single season (1984/85), the test can only exclude large-scale amplitude variability.

Good agreement of the sizes of the amplitudes to within a few millimag is also found when we compare the present results with those published by Shobbrook & Stobie (1974) as well as Balona & Stobie (1980). This result confirms the lack of large amplitude variability in this time period. However, note that Balona et al. (2001) report a decreased amplitude of 0.012 mag for $f_3$ in the year 2000.

## The Odessa data

The Odessa measurements of 1 Mon used in this paper are listed in Table 2.

**Acknowledgments.** It is a pleasure to thank Tatyana Dorokhova for helpful comments. Part of the investigation has been supported by the Austrian Fonds zur Förderung der wissenschaftlichen Forschung, project number P17441-N02.

## References

Balona, L. A., Bartlett, B., Caldwell, J. A. R., et al. 2001, MNRAS, 321, 239
Balona, L. A., & Stobie, R. S. 1980, MNRAS, 190, 931
Cousins, A. W. J. 1963, MNSSA, 22, 12
Lenz, P., & Breger, M. 2005, CoAst, 146, 53
Fedotov, Yu. T., & Gladyr, A. V. 1991, Astr. Tsirk., 1548, 19
Millis, R. L., 1973, PASP, 85, 410
Romanov, Yu. S., & Fedotov, Yu. T. 1979, Astr. Tsirk., 1071, 6
Shobbrook, R. R., & Stobie, R. S. 1974, MNRAS, 169, 643

Table 2: Odessa measurements of 1 Mon ($V$ filter).

| HJD 2440000+ | $V$ mag | HJD 2440000+ | $V$ mag | HJD 2440000+ | $V$ mag | HJD 2440000+ | $V$ mag |
|---|---|---|---|---|---|---|---|
| 3904.3241 | 6.215 | 3931.3529 | 6.132 | 5264.5493 | 5.994 | 5353.1916 | 6.032 |
| 3904.3336 | 6.265 | 3932.2649 | 6.024 | 5264.5560 | 6.048 | 5353.1990 | 6.088 |
| 3904.3387 | 6.250 | 3932.2669 | 6.025 | 5264.5626 | 6.078 | 5353.2060 | 6.157 |
| 3904.3396 | 6.240 | 3932.2713 | 6.029 | 5264.5701 | 6.097 | 5353.2138 | 6.176 |
| 3904.3467 | 6.240 | 3932.2785 | 6.040 | 5264.5769 | 6.121 | 5353.2227 | 6.207 |
| 3904.3483 | 6.246 | 3932.2794 | 6.053 | 5272.4602 | 6.033 | 5353.2301 | 6.230 |
| 3904.3526 | 6.224 | 3932.2860 | 6.064 | 5272.4689 | 6.026 | 5353.2366 | 6.253 |
| 3904.3535 | 6.208 | 3932.2868 | 6.051 | 5272.4786 | 6.083 | 5353.2435 | 6.250 |
| 3904.3596 | 6.188 | 3932.2912 | 6.073 | 5272.4878 | 6.100 | 5353.2504 | 6.257 |
| 3904.3606 | 6.186 | 3932.2923 | 6.057 | 5272.4968 | 6.170 | 5353.2577 | 6.272 |
| 3904.3696 | 6.138 | 3932.2975 | 6.082 | 5272.5064 | 6.210 | 5353.2650 | 6.247 |
| 3904.3706 | 6.138 | 3932.3019 | 6.069 | 5272.5147 | 6.170 | 5353.2730 | 6.210 |
| 3904.3756 | 6.084 | 3932.3930 | 6.096 | 5272.5234 | 6.165 | 5353.2811 | 6.171 |
| 3904.3805 | 6.074 | 3932.3098 | 6.106 | 5272.5314 | 6.118 | 5353.2984 | 6.103 |
| 3904.3817 | 6.035 | 3932.3150 | 6.116 | 5272.5399 | 6.082 | 5353.2977 | 6.047 |
| 3904.3877 | 6.005 | 3932.3160 | 6.122 | 5272.5464 | 6.059 | 5353.3043 | 6.005 |
| 3904.3887 | 5.992 | 3932.3206 | 6.153 | 5272.5541 | 6.010 | 5353.3101 | 5.983 |
| 3904.3963 | 6.011 | 3932.3217 | 6.145 | 5272.5609 | 6.002 | 5353.3169 | 5.970 |
| 3904.3975 | 5.992 | 3932.3287 | 6.179 | 5272.5720 | 5.971 | 5353.3232 | 5.994 |
| 3904.4023 | 5.977 | 3932.3298 | 6.209 | 5272.5817 | 6.021 | 5353.3291 | 6.030 |
| 3904.4035 | 6.017 | 3932.3352 | 6.226 | 5272.5907 | 6.041 | 5353.3356 | 6.060 |
| 3904.4087 | 6.029 | 3932.3409 | 6.216 | 5272.5986 | 6.081 | 5353.3435 | 6.116 |
| 3904.4098 | 6.031 | 3932.3419 | 6.223 | 5282.4932 | 6.200 | 5353.3494 | 6.141 |
| 3906.3321 | 6.168 | 3935.2467 | 6.246 | 5282.5006 | 6.165 | 5353.3556 | 6.168 |
| 3906.3332 | 6.189 | 3935.2477 | 6.228 | 5282.5086 | 6.125 | 5353.3617 | 6.175 |
| 3906.3407 | 6.233 | 3935.2524 | 6.251 | 5282.5166 | 6.077 | 5353.3687 | 6.206 |
| 3906.3484 | 6.268 | 3935.2580 | 6.210 | 5282.5246 | 6.045 | 5353.3756 | 6.243 |
| 3906.3495 | 6.307 | 3935.2592 | 6.214 | 5282.5369 | 6.027 | 5353.3813 | 6.235 |
| 3906.3558 | 6.304 | 3935.2670 | 6.185 | 5282.5450 | 6.047 | 5353.3870 | 6.272 |
| 3906.3568 | 6.299 | 3935.2681 | 6.151 | 5282.5548 | 6.088 | 5353.3935 | 6.266 |
| 3906.3634 | 6.287 | 3935.2726 | 6.129 | 5282.5616 | 6.098 | 5353.3995 | 6.221 |
| 3906.3644 | 6.304 | 3935.2735 | 6.113 | 5282.5717 | 6.163 | 6006.5168 | 6.251 |
| 3906.3720 | 6.295 | 3935.2779 | 6.078 | 5282.5791 | 6.191 | 6006.5190 | 6.234 |
| 3906.3729 | 6.293 | 3935.2791 | 6.070 | 5282.5901 | 6.237 | 6006.5212 | 6.202 |
| 3906.3788 | 6.260 | 3935.2831 | 6.052 | 5282.5969 | 6.242 | 6006.5245 | 6.167 |
| 3906.3797 | 6.249 | 3935.2843 | 6.055 | 5282.6078 | 6.262 | 6006.5267 | 6.164 |
| 3906.3847 | 6.224 | 3935.2911 | 6.004 | 5284.5311 | 6.244 | 6006.5288 | 6.146 |
| 3906.3862 | 6.203 | 3935.2923 | 6.005 | 5284.5384 | 6.171 | 6006.5308 | 6.137 |
| 3931.2586 | 6.232 | 3935.2998 | 5.983 | 5284.5453 | 6.051 | 6006.5336 | 6.130 |
| 3931.2598 | 6.260 | 3935.3010 | 5.989 | 5284.5520 | 6.008 | 6006.5357 | 6.093 |
| 3931.2662 | 6.271 | 3935.3071 | 6.020 | 5284.5586 | 6.000 | 6006.5377 | 6.070 |
| 3931.2719 | 6.240 | 3935.3082 | 6.005 | 5284.5657 | 6.015 | 6006.5399 | 6.062 |
| 3931.2803 | 6.262 | 3935.3135 | 6.041 | 5284.5728 | 6.044 | 6006.5421 | 6.013 |
| 3931.2861 | 6.219 | 3935.3147 | 6.057 | 5284.5807 | 6.098 | 6006.5442 | 6.000 |
| 3931.2922 | 6.138 | 3935.3198 | 6.063 | 5284.5778 | 6.141 | 6006.5463 | 5.985 |
| 3931.2973 | 6.097 | 3935.3209 | 6.060 | 5285.3590 | 5.989 | 6006.5489 | 5.983 |
| 3931.3055 | 6.007 | 3935.3253 | 6.106 | 5285.3663 | 5.975 | 6006.5512 | 5.966 |
| 3931.3072 | 6.013 | 3935.3300 | 6.097 | 5285.3745 | 6.033 | 6006.5532 | 5.955 |
| 3931.3130 | 5.992 | 3935.3335 | 6.102 | 5285.3828 | 6.097 | 6006.5553 | 5.962 |
| 3931.3142 | 6.003 | 3935.3344 | 6.152 | 5285.3905 | 6.095 | 6006.5573 | 5.945 |
| 3931.3196 | 5.989 | 5264.4980 | 6.123 | 5285.4000 | 6.104 | 6006.5593 | 5.950 |
| 3931.3270 | 5.982 | 5264.5076 | 6.091 | 5285.4080 | 6.146 | 6006.5614 | 5.962 |
| 3931.3318 | 5.998 | 5264.5144 | 6.062 | 5285.4163 | 6.217 | 6006.5635 | 5.967 |
| 3931.3366 | 6.033 | 5264.5220 | 6.019 | 5285.4364 | 6.307 | 6006.5665 | 6.000 |
| 3931.3452 | 6.097 | 5264.5305 | 6.010 | 5285.4475 | 6.284 | 6006.5686 | 5.995 |
| 3931.3498 | 6.113 | 5264.5416 | 5.985 | 5353.1845 | 6.019 | 6006.5708 | 6.004 |

Table 2 continued.

| | | | | | | | |
|---|---|---|---|---|---|---|---|
| 6006.5728 | 6.028 | 6038.3726 | 6.167 | 6039.3526 | 6.035 | 6039.4917 | 6.075 |
| 6006.6037 | 6.183 | 6038.3751 | 6.159 | 6039.3547 | 6.008 | 6039.4937 | 6.061 |
| 6006.6058 | 6.194 | 6038.3771 | 6.129 | 6039.3574 | 6.013 | 6039.4973 | 6.055 |
| 6006.6080 | 6.196 | 6038.3791 | 6.149 | 6039.3597 | 6.039 | 6039.4993 | 6.061 |
| 6006.6108 | 6.213 | 6038.3811 | 6.141 | 6039.3629 | 6.037 | 6039.5014 | 6.058 |
| 6006.6128 | 6.215 | 6038.3837 | 6.111 | 6039.3650 | 6.025 | 6039.5034 | 6.051 |
| 6006.6149 | 6.253 | 6038.3839 | 6.102 | 6039.3768 | 6.059 | 6039.5055 | 6.052 |
| 6030.4579 | 6.244 | 6038.3885 | 6.071 | 6039.3798 | 6.086 | 6039.5086 | 6.071 |
| 6030.4601 | 6.240 | 6038.3905 | 6.050 | 6039.3818 | 6.090 | 6039.5106 | 6.065 |
| 6030.4623 | 6.243 | 6038.3924 | 6.040 | 6039.3840 | 6.095 | 6039.5131 | 6.075 |
| 6030.4645 | 6.230 | 6038.3943 | 6.030 | 6039.3860 | 6.111 | 6039.5151 | 6.076 |
| 6030.4677 | 6.229 | 6038.3964 | 6.025 | 6039.3889 | 6.122 | 6039.5184 | 6.087 |
| 6030.4700 | 6.210 | 6038.3984 | 6.005 | 6039.3909 | 6.121 | 6039.5206 | 6.097 |
| 6030.4725 | 6.211 | 6038.4010 | 6.000 | 6039.3930 | 6.118 | 6039.5231 | 6.114 |
| 6030.4757 | 6.192 | 6038.4030 | 6.019 | 6039.3959 | 6.161 | 6039.5252 | 6.119 |
| 6030.4781 | 6.186 | 6038.4050 | 5.983 | 6039.3980 | 6.163 | 6039.5284 | 6.119 |
| 6030.4805 | 6.159 | 6038.4070 | 5.997 | 6039.4001 | 6.150 | 6039.5305 | 6.137 |
| 6030.4840 | 6.122 | 6038.4094 | 5.998 | 6039.4022 | 6.156 | 6039.5326 | 6.145 |
| 6030.4862 | 6.111 | 6038.4115 | 6.005 | 6039.4053 | 6.153 | 6039.5352 | 6.138 |
| 6030.4888 | 6.086 | 6038.4141 | 5.995 | 6039.4074 | 6.160 | 6039.5375 | 6.138 |
| 6030.4910 | 6.070 | 6038.4161 | 6.008 | 6039.4095 | 6.185 | 6039.5404 | 6.153 |
| 6030.4934 | 6.014 | 6038.4182 | 6.027 | 6039.4116 | 6.189 | 6039.5428 | 6.168 |
| 6030.4992 | 5.992 | 6038.4204 | 6.031 | 6039.4138 | 6.188 | 6039.5449 | 6.159 |
| 6030.5015 | 5.999 | 6038.4224 | 6.040 | 6039.4167 | 6.176 | 6039.5477 | 6.168 |
| 6030.5037 | 5.976 | 6038.4244 | 6.067 | 6039.4188 | 6.187 | 6039.5499 | 6.183 |
| 6030.5078 | 5.978 | 6038.4268 | 6.062 | 6039.4208 | 6.196 | 6040.3950 | 6.174 |
| 6030.5109 | 5.990 | 6038.4290 | 6.109 | 6039.4230 | 6.193 | 6040.3975 | 6.175 |
| 6030.5130 | 5.984 | 6038.4321 | 6.097 | 6039.4252 | 6.207 | 6040.3994 | 6.183 |
| 6030.5153 | 5.988 | 6038.4343 | 6.127 | 6039.4279 | 6.214 | 6040.4019 | 6.187 |
| 6030.5181 | 6.009 | 6038.4365 | 6.133 | 6039.4299 | 6.219 | 6040.4052 | 6.172 |
| 6030.5204 | 6.010 | 6038.4385 | 6.126 | 6039.4320 | 6.206 | 6040.4071 | 6.199 |
| 6030.5234 | 6.022 | 6038.4407 | 6.145 | 6039.4341 | 6.201 | 6040.4078 | 6.180 |
| 6030.5256 | 6.042 | 6038.4432 | 6.163 | 6039.4364 | 6.212 | 6040.4094 | 6.193 |
| 6030.5276 | 6.035 | 6038.4459 | 6.165 | 6039.4386 | 6.198 | 6040.4104 | 6.193 |
| 6030.5303 | 6.057 | 6038.4479 | 6.181 | 6039.4425 | 6.199 | 6040.4119 | 6.186 |
| 6030.5326 | 6.064 | 6038.4499 | 6.189 | 6039.4445 | 6.190 | 6040.4130 | 6.182 |
| 6030.5347 | 6.072 | 6038.4520 | 6.192 | 6039.4465 | 6.201 | 6040.4146 | 6.189 |
| 6030.5368 | 6.074 | 6038.4541 | 6.205 | 6039.4487 | 6.196 | 6040.4174 | 6.177 |
| 6030.5397 | 6.100 | 6038.4561 | 6.214 | 6039.4507 | 6.186 | 6040.4185 | 6.176 |
| 6030.5417 | 6.117 | 6038.4581 | 6.201 | 6039.4528 | 6.195 | 6040.4204 | 6.175 |
| 6030.5438 | 6.114 | 6038.4603 | 6.230 | 6039.4551 | 6.174 | 6040.4225 | 6.170 |
| 6030.5471 | 6.112 | 6038.4633 | 6.223 | 6039.4583 | 6.175 | 6040.4235 | 6.164 |
| 6030.5494 | 6.135 | 6038.4655 | 6.233 | 6039.4605 | 6.164 | 6040.4257 | 6.166 |
| 6030.5516 | 6.131 | 6038.4676 | 6.229 | 6039.4624 | 6.171 | 6040.4276 | 6.171 |
| 6030.5548 | 6.155 | 6038.4698 | 6.218 | 6039.4645 | 6.151 | 6040.4295 | 6.161 |
| 6030.5570 | 6.162 | 6038.4719 | 6.225 | 6039.4677 | 6.144 | 6040.4324 | 6.161 |
| 6030.5592 | 6.180 | 6038.4740 | 6.229 | 6039.4697 | 6.128 | 6040.4344 | 6.149 |
| 6030.5641 | 6.198 | 6038.4762 | 6.232 | 6039.4717 | 6.112 | 6040.4362 | 6.148 |
| 6030.5669 | 6.205 | 6038.4789 | 6.237 | 6039.4737 | 6.109 | 6040.4381 | 6.126 |
| 6030.5692 | 6.212 | 6038.4810 | 6.234 | 6039.4758 | 6.109 | 6040.4401 | 6.139 |
| 6030.5714 | 6.207 | 6038.4832 | 6.239 | 6039.4780 | 6.098 | 6040.4420 | 6.125 |
| 6030.5744 | 6.219 | 6038.4859 | 6.228 | 6039.4811 | 6.098 | 6040.4456 | 6.124 |
| 6030.5777 | 6.222 | 6038.4880 | 6.247 | 6039.4830 | 6.105 | 6040.4475 | 6.115 |
| 6030.5800 | 6.225 | 6039.3454 | 6.066 | 6039.4850 | 6.077 | 6040.4494 | 6.113 |
| 6038.3688 | 6.186 | 6039.3482 | 6.061 | 6039.4871 | 6.077 | 6040.4513 | 6.102 |
| 6038.3707 | 6.191 | 6039.3504 | 6.053 | 6039.4892 | 6.077 | 6040.4532 | 6.109 |

Table 2 continued.

| | | | | | | | |
|---|---|---|---|---|---|---|---|
| 6040.4560 | 6.092 | 6114.2769 | 6.142 | 6118.2548 | 6.127 | 6119.3115 | 6.020 |
| 6040.4584 | 6.085 | 6114.2792 | 6.161 | 6118.2579 | 6.131 | 6119.3141 | 5.998 |
| 6040.4622 | 6.075 | 6114.2826 | 6.150 | 6118.2598 | 6.141 | 6119.3177 | 6.051 |
| 6040.4642 | 6.069 | 6114.2838 | 6.144 | 6118.2619 | 6.139 | 6119.3203 | 6.071 |
| 6040.4654 | 6.068 | 6114.2852 | 6.180 | 6118.2713 | 6.171 | 6119.3234 | 6.069 |
| 6040.4673 | 6.073 | 6114.2875 | 6.188 | 6118.2738 | 6.163 | 6119.3246 | 6.062 |
| 6040.4692 | 6.066 | 6115.2948 | 6.094 | 6118.2762 | 6.166 | 6119.3281 | 6.108 |
| 6040.4714 | 6.062 | 6115.2977 | 6.088 | 6118.2796 | 6.158 | 6119.3314 | 6.119 |
| 6040.4733 | 6.059 | 6115.3011 | 6.061 | 6118.2819 | 6.172 | 6119.3348 | 6.130 |
| 6040.4763 | 6.059 | 6115.3033 | 6.044 | 6118.2841 | 6.165 | 6119.3374 | 6.112 |
| 6040.4783 | 6.077 | 6115.3061 | 6.051 | 6118.2864 | 6.182 | 6119.3402 | 6.161 |
| 6040.4808 | 6.071 | 6115.3087 | 6.025 | 6118.2885 | 6.179 | 6120.2011 | 6.187 |
| 6040.4843 | 6.066 | 6115.3118 | 6.015 | 6118.2918 | 6.163 | 6120.2048 | 6.174 |
| 6040.4863 | 6.092 | 6115.3141 | 6.008 | 6118.2941 | 6.162 | 6120.2078 | 6.161 |
| 6040.4887 | 6.084 | 6115.3224 | 6.001 | 6118.2970 | 6.163 | 6120.2109 | 6.137 |
| 6040.4909 | 6.092 | 6115.3246 | 5.976 | 6118.2997 | 6.150 | 6120.2139 | 6.103 |
| 6040.4930 | 6.097 | 6115.3269 | 5.969 | 6118.3028 | 6.134 | 6120.2167 | 6.081 |
| 6040.4950 | 6.103 | 6115.3293 | 5.999 | 6118.3060 | 6.124 | 6120.2177 | 6.064 |
| 6040.4972 | 6.100 | 6115.3317 | 5.990 | 6118.3083 | 6.116 | 6120.2197 | 6.069 |
| 6040.4998 | 6.113 | 6115.3350 | 6.038 | 6118.3112 | 6.103 | 6120.2204 | 6.053 |
| 6040.5029 | 6.126 | 6115.3375 | 6.061 | 6118.3135 | 6.098 | 6120.2236 | 6.006 |
| 6040.5056 | 6.130 | 6115.3399 | 6.074 | 6118.3173 | 6.102 | 6120.2263 | 6.006 |
| 6040.5082 | 6.120 | 6115.3421 | 6.058 | 6118.3198 | 6.084 | 6120.2285 | 5.985 |
| 6040.5120 | 6.138 | 6115.3448 | 6.093 | 6118.3221 | 6.090 | 6120.2313 | 5.970 |
| 6040.5147 | 6.147 | 6115.3475 | 6.100 | 6118.3245 | 6.073 | 6120.2335 | 5.958 |
| 6040.5180 | 6.139 | 6115.3505 | 6.136 | 6118.3274 | 6.073 | 6120.2363 | 5.964 |
| 6040.5200 | 6.154 | 6115.3532 | 6.152 | 6118.3310 | 6.064 | 6120.2389 | 5.916 |
| 6040.5221 | 6.170 | 6115.3560 | 6.142 | 6118.3334 | 6.065 | 6120.2415 | 5.929 |
| 6114.1979 | 6.198 | 6115.3571 | 6.142 | 6118.3359 | 6.043 | 6120.2437 | 5.932 |
| 6114.2009 | 6.166 | 6115.3594 | 6.187 | 6118.3381 | 6.043 | 6120.2465 | 5.917 |
| 6114.2048 | 6.158 | 6115.3619 | 6.180 | 6118.3404 | 6.030 | 6120.2486 | 5.925 |
| 6114.2071 | 6.110 | 6118.1959 | 6.002 | 6118.3439 | 6.033 | 6120.2508 | 5.924 |
| 6114.2111 | 6.080 | 6118.1990 | 6.010 | 6118.3449 | 6.023 | 6120.2539 | 5.990 |
| 6114.2138 | 6.058 | 6118.2022 | 6.017 | 6118.3476 | 6.032 | 6120.2563 | 5.966 |
| 6114.2161 | 6.034 | 6118.2066 | 6.012 | 6118.3500 | 6.030 | 6120.2572 | 5.985 |
| 6114.2205 | 5.987 | 6118.2091 | 6.019 | 6118.3525 | 6.024 | 6120.2590 | 5.980 |
| 6114.2243 | 5.964 | 6118.2114 | 6.016 | 6118.3536 | 6.030 | 6120.2615 | 5.998 |
| 6114.2268 | 5.932 | 6118.2144 | 6.016 | 6119.2619 | 6.153 | 6120.2643 | 6.017 |
| 6114.2306 | 5.914 | 6118.2166 | 6.037 | 6119.2648 | 6.151 | 6120.2665 | 6.000 |
| 6114.2335 | 5.925 | 6118.2195 | 6.026 | 6119.2680 | 6.102 | 6120.2687 | 6.029 |
| 6114.2364 | 5.953 | 6118.2207 | 6.019 | 6119.2686 | 6.113 | 6120.2714 | 6.026 |
| 6114.2373 | 5.954 | 6118.2236 | 6.040 | 6119.2711 | 6.092 | 6120.2738 | 6.053 |
| 6114.2393 | 5.951 | 6118.2259 | 6.041 | 6119.2733 | 6.077 | 6120.2759 | 6.078 |
| 6114.2436 | 5.984 | 6118.2269 | 6.060 | 6119.2762 | 6.052 | 6120.2793 | 6.109 |
| 6114.2462 | 5.984 | 6118.2288 | 6.063 | 6119.2787 | 6.023 | 6120.2819 | 6.106 |
| 6114.2484 | 5.992 | 6118.2317 | 6.056 | 6119.2822 | 5.995 | 6120.2840 | 6.143 |
| 6114.2520 | 6.012 | 6118.2342 | 6.065 | 6119.2847 | 5.985 | 6120.2861 | 6.144 |
| 6114.2542 | 6.041 | 6118.2365 | 6.079 | 6119.2871 | 5.988 | 6123.2045 | 5.979 |
| 6114.2565 | 6.043 | 6118.2374 | 6.091 | 6119.2904 | 5.992 | 6123.2069 | 5.972 |
| 6114.2579 | 6.049 | 6118.2401 | 6.086 | 6119.2934 | 5.987 | 6123.2095 | 5.956 |
| 6114.2605 | 6.073 | 6118.2424 | 6.094 | 6119.2957 | 5.961 | 6123.2119 | 5.973 |
| 6114.2646 | 6.086 | 6118.2435 | 6.093 | 6119.2987 | 5.974 | 6123.2141 | 5.965 |
| 6114.2656 | 6.075 | 6118.2454 | 6.086 | 6119.3011 | 5.982 | 6123.2164 | 5.960 |
| 6114.2680 | 6.105 | 6118.2484 | 6.101 | 6119.3035 | 5.972 | 6123.2184 | 5.976 |
| 6114.2706 | 6.127 | 6118.2504 | 6.114 | 6119.3065 | 6.011 | 6123.2210 | 5.965 |
| 6114.2745 | 6.154 | 6118.2525 | 6.125 | 6119.3092 | 6.001 | 6123.2233 | 5.989 |

Table 2 continued.

| | | | | | | | |
|---|---|---|---|---|---|---|---|
| 6123.2254 | 6.010 | 6125.2282 | 6.078 | 6126.2122 | 6.045 | 6126.2918 | 6.128 |
| 6123.2275 | 5.994 | 6125.2308 | 6.102 | 6126.2144 | 6.061 | 6126.2940 | 6.128 |
| 6123.2303 | 6.029 | 6125.2347 | 6.119 | 6126.2171 | 6.052 | 6126.2962 | 6.127 |
| 6123.2323 | 6.027 | 6125.2373 | 6.100 | 6126.2192 | 6.023 | 6132.2021 | 6.041 |
| 6123.2355 | 6.059 | 6125.2393 | 6.071 | 6126.2219 | 6.008 | 6132.2042 | 6.032 |
| 6123.2384 | 6.060 | 6125.2424 | 6.069 | 6126.2239 | 6.020 | 6132.2065 | 6.037 |
| 6123.2406 | 6.086 | 6125.2445 | 6.066 | 6126.2263 | 5.995 | 6132.2088 | 6.043 |
| 6123.2433 | 6.081 | 6125.2471 | 6.068 | 6126.2303 | 5.989 | 6132.2112 | 6.040 |
| 6123.2454 | 6.098 | 6125.2492 | 6.063 | 6126.2310 | 5.999 | 6132.2138 | 6.064 |
| 6123.2474 | 6.085 | 6125.2513 | 6.044 | 6126.2333 | 5.985 | 6132.2161 | 6.085 |
| 6123.2495 | 6.110 | 6125.2534 | 6.039 | 6126.2354 | 6.012 | 6132.2182 | 6.086 |
| 6123.2522 | 6.131 | 6125.2562 | 6.016 | 6126.2385 | 6.018 | 6132.2208 | 6.084 |
| 6123.2543 | 6.136 | 6125.2584 | 6.039 | 6126.2412 | 6.012 | 6132.2227 | 6.106 |
| 6123.2564 | 6.138 | 6125.2606 | 6.017 | 6126.2432 | 6.013 | 6132.2252 | 6.112 |
| 6123.2584 | 6.146 | 6125.2634 | 6.017 | 6126.2454 | 5.987 | 6132.2277 | 6.107 |
| 6123.2611 | 6.160 | 6125.2661 | 6.021 | 6126.2482 | 6.000 | 6132.2301 | 6.130 |
| 6123.2633 | 6.181 | 6125.2688 | 6.033 | 6126.2503 | 6.023 | 6132.2324 | 6.139 |
| 6123.2655 | 6.173 | 6125.2709 | 6.040 | 6126.2532 | 6.027 | 6132.2346 | 6.140 |
| 6123.2684 | 6.170 | 6125.2731 | 6.028 | 6126.2555 | 6.044 | 6132.2369 | 6.144 |
| 6123.2705 | 6.195 | 6125.2759 | 6.028 | 6126.2577 | 6.040 | 6132.2398 | 6.151 |
| 6123.2725 | 6.190 | 6125.2783 | 6.023 | 6126.2605 | 6.044 | 6132.2424 | 6.151 |
| 6123.2746 | 6.181 | 6125.2802 | 6.016 | 6126.2628 | 6.061 | 6132.2448 | 6.177 |
| 6123.2774 | 6.197 | 6125.2833 | 6.042 | 6126.2648 | 6.056 | 6132.2471 | 6.182 |
| 6123.2796 | 6.198 | 6125.2852 | 6.049 | 6126.2676 | 6.073 | 6132.2495 | 6.179 |
| 6123.2824 | 6.206 | 6125.2873 | 6.074 | 6126.2725 | 6.080 | 6132.2519 | 6.179 |
| 6123.2850 | 6.205 | 6125.2894 | 6.066 | 6126.2747 | 6.107 | 6132.2543 | 6.184 |
| 6123.2882 | 6.203 | 6125.2915 | 6.060 | 6126.2770 | 6.113 | 6132.2567 | 6.189 |
| 6123.2903 | 6.222 | 6125.2943 | 6.081 | 6126.2791 | 6.121 | 6132.2595 | 6.205 |
| 6123.2932 | 6.220 | 6125.2968 | 6.087 | 6126.2820 | 6.111 | 6132.2619 | 6.199 |
| 6123.2955 | 6.234 | 6125.2995 | 6.085 | 6126.2842 | 6.120 | 6132.2642 | 6.180 |
| 6125.2230 | 6.105 | 6126.2032 | 6.106 | 6126.2866 | 6.121 | 6132.2670 | 6.197 |
| 6125.2261 | 6.111 | 6126.2090 | 6.061 | 6126.2888 | 6.134 | | |

*Comm. in Asteroseismology*
*Vol. 148, 2006*

# Amplitude variability of the cepheid V473 Lyr

M. Breger

Institut für Astronomie, Türkenschanzstrasse 17, 1180 Vienna, Austria

## Abstract

The star V473 Lyr is a short-period cepheid (P = 1.49d) with a very strong modulation of the amplitude on a time scale near 1200d. Previously unpublished photometry is presented. This includes good coverage near the maximum amplitude. The available photometry and radial-velocity measurements going back to 1966 are analyzed with PERIOD04 to examine whether the amplitude variability is caused by beating between close frequencies. In particular, we have looked for the existence of a close frequency triplet with equidistant frequency spacing. This is required by the Combination Mode Hypothesis, which explains equidistant frequency triplets by a combination of two frequencies. The frequency triplet determined by us for V473 Lyr misses equidistancy by 0.00005 $\pm$ 0.00001 cd$^{-1}$.

## Introduction

V473 Lyr (HR 7308) is probably the best-studied cepheid with a strong Blazhko Effect. This 1.49d cepheid has a modulation period near 1200d. Breger (1981) as well as Burki, Mayor & Benz (1982) have shown that a simple model based on beating between two closely spaced frequencies cannot explain the observed phase and amplitude variability. Furthermore, the long modulation period rules out rotational splitting of nonradial modes as well.

The purpose of this paper is to examine the amplitude and phase variability of this cepheid. In particular, we are interested in testing the Combination Mode Hypothesis, which could explain equidistant frequency triplets seen in different types of pulsating stars (Breger & Kolenberg 2006). Here, the third frequency, $f_3$ would be related to the combination of the other two modes of higher amplitude by $f_3 = 2f_1 - f_2$. A necessary condition is that the three modes are equidistant (or nearly so, if another mode is excited by resonance).

An equidistant or near-equidistant frequency triplet has been reported for V473 Lyr: Koen (2001) analyzed the available Hipparcos photometry and proposed that the amplitude variations of V473 Lyr could be described by symmetrical frequency triplets separated by 0.0011 c/d. His analysis of the Burki, Mayor & Benz (1982) radial-velocity data showed a small departure from equidistancy: 0.0008 vs 0.0009 c/d.The star, therefore, is an interesting candidate for further investigation.

## Previously unpublished photometric data

During 1980 and 1981, photometry of V473 Lyr was obtained with the 0.9-m telescope at McDonald Observatory. The measurements used the $V$ filter and the observers were led by the present author. The measurements relative to the comparison star, HR 7280, are listed in Table 1.

One of the aims of these observations was to measure the night-to-night variations of the cepheid light curves near the Blazhko phase of maximum amplitude. A few closely spaced nights are shown in Fig. 1, which confirm the rapid change from night to night of the light curve near maximum. Note that the 1.49d period allows a comparison every three nights.

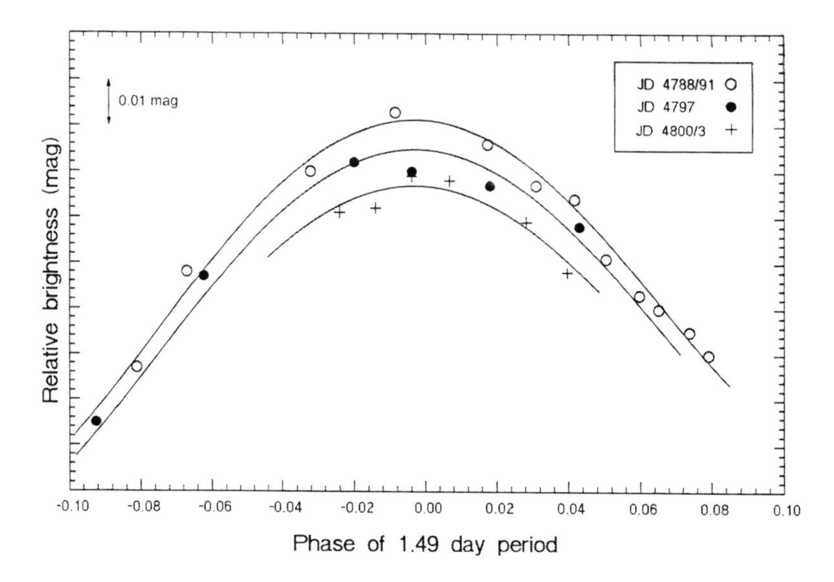


Figure 1: Rapid amplitude variations near maximum light. The times are listed relative to HJD 244 0000.

Table 1: Photometric measurements of V473 Lyr ($V$ filter)

| HJD 2440000+ | $\delta V$ mag | HJD 2440000+ | $\delta V$ mag | HJD 2440000+ | $\delta V$ mag | HJD 2440000+ | $\delta V$ mag |
|---|---|---|---|---|---|---|---|
| 4416.7562 | -0.133 | 4788.8760 | -0.368 | 4798.6450 | -0.006 | 4800.8910 | -0.340 |
| 4416.7686 | -0.129 | 4791.8120 | -0.361 | 4798.6580 | -0.007 | 4800.9050 | -0.328 |
| 4439.6596 | -0.162 | 4791.8200 | -0.363 | 4798.6730 | -0.009 | 4803.6550 | -0.327 |
| 4439.7259 | -0.188 | 4791.8270 | -0.362 | 4798.6830 | -0.004 | 4803.6680 | -0.337 |
| 4439.7931 | -0.211 | 4791.8350 | -0.373 | 4798.7310 | 0.005 | 4803.6760 | -0.348 |
| 4440.6466 | -0.124 | 4791.8420 | -0.369 | 4798.7420 | -0.006 | 4803.7200 | -0.368 |
| 4440.7220 | -0.115 | 4791.8490 | -0.367 | 4798.7560 | -0.008 | 4803.7280 | -0.370 |
| 4440.7972 | -0.109 | 4791.8600 | -0.370 | 4798.7760 | -0.006 | 4803.7550 | -0.379 |
| 4440.8670 | -0.108 | 4791.8900 | -0.359 | 4798.7900 | -0.002 | 4803.7630 | -0.383 |
| 4440.8939 | -0.110 | 4791.9000 | -0.355 | 4798.8000 | -0.002 | 4803.7820 | -0.380 |
| 4441.6518 | -0.238 | 4791.9070 | -0.348 | 4798.8080 | -0.005 | 4803.8380 | -0.364 |
| 4441.6961 | -0.233 | 4791.9220 | -0.342 | 4798.8180 | -0.007 | 4803.8510 | -0.351 |
| 4441.7393 | -0.222 | 4796.6810 | -0.188 | 4798.8270 | -0.016 | 4803.8700 | -0.342 |
| 4441.7785 | -0.215 | 4796.7140 | -0.176 | 4798.8470 | -0.007 | 4803.8780 | -0.333 |
| 4441.8059 | -0.206 | 4796.7230 | -0.176 | 4798.8610 | -0.017 | 4803.8860 | -0.335 |
| 4441.8362 | -0.196 | 4796.7330 | -0.164 | 4798.8750 | -0.032 | 4803.8940 | -0.330 |
| 4441.8691 | -0.190 | 4796.7430 | -0.156 | 4798.8850 | -0.032 | 4803.9170 | -0.314 |
| 4445.6707 | -0.181 | 4796.7550 | -0.141 | 4798.9010 | -0.028 | 4805.6470 | -0.180 |
| 4445.7266 | -0.199 | 4796.7690 | -0.141 | 4798.9100 | -0.035 | 4805.7230 | -0.139 |
| 4445.7469 | -0.209 | 4797.6280 | -0.262 | 4799.6690 | -0.182 | 4805.7350 | -0.132 |
| 4445.7738 | -0.219 | 4797.6420 | -0.287 | 4799.6820 | -0.173 | 4805.7480 | -0.126 |
| 4445.7995 | -0.225 | 4797.6560 | -0.295 | 4799.7190 | -0.155 | 4805.7560 | -0.117 |
| 4445.8228 | -0.231 | 4797.6740 | -0.315 | 4799.7290 | -0.147 | 4805.7710 | -0.118 |
| 4445.8660 | -0.240 | 4797.6870 | -0.317 | 4799.7370 | -0.146 | 4805.7820 | -0.114 |
| 4786.8250 | -0.010 | 4797.7190 | -0.347 | 4799.7460 | -0.146 | 4805.7900 | -0.110 |
| 4786.9010 | -0.013 | 4797.7370 | -0.350 | 4799.7590 | -0.145 | 4805.8030 | -0.106 |
| 4786.9080 | -0.020 | 4797.7680 | -0.358 | 4799.7950 | -0.123 | 4805.8150 | -0.092 |
| 4786.9170 | -0.013 | 4797.7820 | -0.372 | 4799.8160 | -0.122 | 4805.8310 | -0.089 |
| 4787.7690 | -0.171 | 4797.7920 | -0.375 | 4800.8200 | -0.376 | 4805.8400 | -0.081 |
| 4787.7830 | -0.158 | 4797.8060 | -0.370 | 4800.8270 | -0.373 | 4805.8480 | -0.079 |
| 4787.8040 | -0.150 | 4797.8240 | -0.371 | 4800.8400 | -0.367 | 4805.8570 | -0.073 |
| 4787.8170 | -0.144 | 4797.8390 | -0.367 | 4800.8480 | -0.363 | 4805.8790 | -0.069 |
| 4787.8230 | -0.146 | 4797.8580 | -0.367 | 4800.8620 | -0.355 | 4805.8910 | -0.066 |
| 4787.8320 | -0.143 | 4797.8760 | -0.358 | 4800.8750 | -0.347 | 4805.9090 | -0.058 |
| 4787.9340 | -0.084 | 4797.8870 | -0.355 | 4800.8830 | -0.343 | 4805.9170 | -0.051 |

## Multifrequency analysis

We have collected the available photometric data from 1966 to 1983 (Percy & Evans 1980, Breger 1981, Henrikssen 1983, Burki et al. 1986) as well as photometry in Table 1. The periodicities in the data were analyzed with the PERIOD04 statistical package (Lenz & Breger 2005). We ignore complications such as the increase in asymmetry of the light curve shape with increasing amplitude. Not surprisingly, the previously known main frequency of 0.67075 $cd^{-1}$ is correctly determined. However, in the Fourier spectrum many additional peaks in the vicinity of the main peak are present (before and after prewhitening the main frequency) and no unique multifrequency solution can be obtained. This failure is caused in part by the long modulation period of $\sim$ 3 years together with a spotty coverage (e.g., annual aliasing).

Next, we added the excellent radial-velocity data by Burki, Mayor & Benz (1982). Since the radial-velocity variations resemble inverted light curves, we have converted the radial-velocity variations to light variations, using the phase differences and amplitude ratios determined for this star by Burki et al. (1986) in their Table 4. The results are now improved: the larger data set revealed three close main frequencies with their harmonics and a combination. This is shown in Table 2.

The multifrequency analysis confirmed the existence of the frequency triplet ($f_1$ to $f_3$ in Table 2). The frequency triplet determined by us misses equidistancy by $0.00005 \pm 0.00001$ $cd^{-1}$. Because of the small formal error the result appears to be statistically significant. However, the solution may be less certain for the following reasons:

(i) Examination of the observed and fitted light curves reveals that the fits are not good: the standard deviation is 0.023 mag per single measurement.

(ii) Even after the inclusion of the radial-velocity data, the spectral window indicates that the frequencies and their harmonics are not statistically independent of each other. Note here the unfavorable main pulsation period of 1.49 d. This represents a severe problem.

(iii) Furthermore, analysis of the residuals shows additional periodicities near the main frequencies.

We conclude that the data for the best-studied Blazhko cepheid are insufficient to test the Combination Mode Hypothesis.

**Acknowledgments.** Part of the investigation has been supported by the Austrian Fonds zur Förderung der wissenschaftlichen Forschung.

Table 2: Multiple frequencies of V473 Lyr

| Frequency | $cd^{-1}$ | Amplitude in $V$ mag |
|---|---|---|
| $f_1$ | 0.67075 | 0.0764 |
| $f_2$ | 0.67157 | 0.0354 |
| $f_3$ | 0.66988 | 0.0292 |
| $2f_1$ | 1.34150 | 0.0132 |
| $2f_2$ | 1.34313 | 0.0061 |
| $2f_3$ | 1.33977 | 0.0081 |
| $3f_1$ | 2.01226 | 0.0023 |
| $3f_2$ | 2.01470 | 0.0025 |
| $3f_3$ | 2.00965 | 0.0004 |
| $f_1+f_2$ | 1.34232 | 0.0110 |

## References

Breger, M. 1981, ApJ, 249, 666
Breger, M., & Kolenberg, K. 2006, A&A, in press
Burki G., Mayor M., & Benz W. 1982, A&A, 109, 258
Burki, G., Schmidt, E. G., Aranello Ferro, A., et al. 1986, A&A, 168, 139
Henrikssen, G. 1983, Uppsala Astr. Obs. Report, 26
Koen, C. 2001, MNRAS, 322, 97
Lenz, P., & Breger, M. 2005, CoAst, 146, 53
Percy, J. R., & Evans, N. R. 1980, AJ, 85, 1509

*Comm. in Asteroseismology*
*Vol. 148, 2006*

# *MOST* Photometry of the roAp star HD 134214

C. Cameron[1], J.M. Matthews[1], J.F. Rowe[1], R. Kuschnig[1], D.B. Guenther[2], A.F.J. Moffat[3], S.M. Rucinski[4], D. Sasselov[5], G.A.H., Walker[6], W.W. Weiss[7]

[1]Dept. of Physics and Astronomy, UBC, 6224 Agricultural Road, Vancouver, BC, V6T 1Z1, Canada
[2]Dept. of Astronomy and Physics, Saint Mary's University, Halifax, NS, B3H 3C3, Canada
[3]Dépt. de physique, Univ. de Montréal C.P. 6128, Succ. Centre-Ville, Montréal, QC H3C 3J7, Canada; and Obs. du Mont Mégantic
[4]Dept. of Astronomy & Astrophysics, David Dunlap Obs., Univ. Toronto P.O. Box 360, Richmond Hill, ON L4C 4Y6, Canada
[5]Harvard-Smithsonian Center for Astrophysics, 60 Garden Street, Cambridge, MA 02138, USA
[6]1234 Hewlett Place, Victoria, BC V8S 4P7, Canada
[7]Institut für Astronomie, Universität Wien Türkenschanzstrasse 17, A–1180 Wien, Austria

## Abstract

We present 10.27 hrs of photometry of the roAp star HD 134214 obtained by the *MOST*[1] satellite. The star is shown to be monoperiodic and oscillating at a frequency of 2948.97 $\pm$ 0.55 $\mu$Hz. This is consistent with earlier ground based photometric campaigns (e.g. Kreidl et al. 1994). We do not detect any of the additional frequencies identified in the recent spectroscopic study by Kurtz et al. (2006) down to an amplitude limit of 0.36 mmag ($2\sigma$ significance limit).

## Introduction

The rapidly oscillating Ap stars (roAp) represent a unique subset of the chemically peculiar stars of the upper main sequence. In general, the Ap stars have

[1]*MOST* (Microvariability and Oscillations of STars) is a Canadian Space Agency mission, operated jointly by Dynacon, Inc., and the Universities of Toronto and British Columbia, with assistance from the University of Vienna.

globally organised magnetic fields of strengths of order kiloGauss, and spectral anomalies that are interpreted as vertical and horizontal chemical inhomogeneities in the stellar atmosphere. The roAp stars (among the coolest members of the Ap class) exhibit rapid oscillations in photometry and spectroscopy. These variations (first observed by Kurtz (1982)) have periods from about 5 to 20 minutes and low amplitudes (B $\lesssim$ 10 mmag). They are consistent with acoustic (p-mode) pulsations of low degree and high radial overtone. A thorough review of the roAp stars is provided by Kurtz & Martinez (2000).

HD 134214 is in many ways a typical Ap star. It has an effective temperature of $\sim$ 7500 K, is a moderately slow rotator with a $v \sin i \approx 2.0$ km/s, and has a strong magnetic field with $B_r \approx -2.9$ kG (parameters are derived in Shavrina et al. 2004). However, among the roAp stars, HD 134214 stands alone as the star with the shortest known period ($\sim 5.6$ minutes). This high frequency oscillation is well above the estimated isothermal acoustic cutoff for a typical A-star model (see, e.g., Audard et al. 1998). There is also evidence that this oscillation frequency is variable at the level of a few tenths of $\mu$Hz over a time scale of approximately one year, even though the amplitude remained stable over many years (see Kreidl et al. 1994). Kurtz (1995) discuss the frequency variability observed in this and other roAp stars.

Recently, Kurtz et al. (2006) have shown that in some roAp stars the spectroscopic variability is dramatically different from that observed photometrically. Although HD 134214 is observed to be monoperiodic in photometry over many years (Kreidl et al. 1994), Kurtz et al. present evidence for up to 6 periodicities in their (admittedly short) spectroscopic time-series of this star.

In this paper we present *MOST* (Microvariability & Oscillations of STars) spacebased photometry of HD 134214 and compare the observed oscillations to the results of Kurtz et al. (2006).

## Photometry and Frequency Analysis

*MOST* is a Canadian microsatellite that was launched into a 820-km Sun-synchronous polar orbit in June 2003. Its primary science objective is to obtain nearly continuous, ultra-precise photometric measurements of stars for the purposes of asteroseismology. The instrument is a 15-cm Rumak-Maksutov telescope that illuminates a CCD photometer through a custom broadband filter (350-700 nm). Detailed technical information on the instrument is contained in Walker et al. (2003). The first science results for the mission were published by Matthews et al. (2004).

The roAp star HD 134214 was observed by *MOST* as a trial Direct Imaging target on 1 May 2006. Direct Imaging is the observing mode where a defocused star image is projected onto an open area of the Science CCD (see

Rowe et al. 2006a and references therein for more details on different observing modes of *MOST*). The star was in one of two target fields observed during each *MOST* satellite orbit (period = 101.413 min), so data are collected with a duty cycle of only $\sim$ 47%. Exposures are 1.5 sec long, obtained every 10 seconds. There were a total of 505 measurements collected during 10.27 hrs.

Photometric reductions for Direct Imaging targets are described by Rowe et al. (2006b). In addition to these procedures, the data are detrended using a running mean of approximately 15 mins to reduce any remaining low-frequency contributions to stray light from scattered Earthshine which is modulated with the *MOST* satellite orbital period. This detrending does not influence the high-frequency domain of interest for roAp oscillations. The resulting changes in the mean level of the light curve are not large enough to affect the measured amplitude of the oscillation signal, within the uncertainty from the fit (see below).

The reduced photometry is presented in Figure 1. The light curve of HD 134214, shown in the top panel, has a standard deviation of $\sim 1.8$ mmag and a $2\sigma$ standard error of 0.2 mmag. Data subsets labelled A and B are shown in the lower two panels.

Frequency determinations are made using CAPER; see Walker et al. (2005) and Saio et al. (2006). CAPER is an iterative procedure that identifies periodicities in Fourier space and simultaneously fits a set of sinusoids in the time domain. This method of determining oscillation parameters from time-series data follows the philosophy of the popular software packages Period98 (Sperl 1998) and Period04 (Lenz & Breger 2005). Discrete Fourier Transforms (DFTs) are calculated and the fit is subtracted from the data successively, until there is no meaningful change in the fit residuals.

The DFT and spectral window function for the HD 134214 data are shown in Figure 2. There is one periodicity identified at 2948.97 $\pm$ 0.55 $\mu$Hz with a signal-to-noise ratio of 15.4. The noise in the Fourier domain was estimated as the mean of the entire spectrum (0.12 mmag). We see no evidence for other periodicities at a detection threshold of about $2\sigma$ (or $3\times$ the noise; see Kuschnig et al. 1997). The fitted values of amplitude and phase are 1.88 $\pm$ 0.09 mmag and 1.97 $\pm$ 0.39 radians respectively. The uncertainties in the fit parameters are determined using a bootstrap technique (see below). The dark line in Figure 1 shows the fit of this single sinusoid plotted over the observations.

The uncertainties we calculate for time-series parameters derived from non-linear least-squares methods depend on the noise of the data (which may be a combination of instrumental and random processes) and on the time sampling. It is also known that fitted phase and frequency parameters are correlated, leading to underestimated uncertainties in these parameters when calculated from

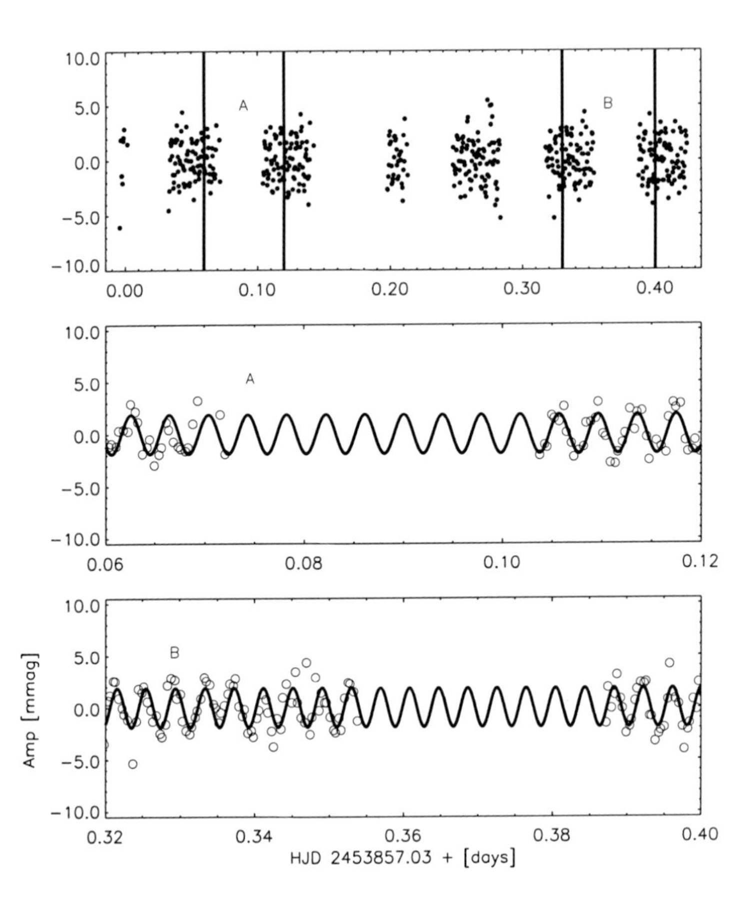


Figure 1: The *MOST* photometry of HD 134214. The top panel shows the 10.27 hrs of data. Gaps in the light curve are the result of observing this target every second satellite orbit. The zoomed regions; labelled A and B, are shown respectively in the lower panels. The open circles are approximately twice the standard error in size ($\sim 0.20$ mmag). The solid line represents the fit of a single sinusoid with a frequency of 2948.97 $\mu$Hz.

a covariance matrix (e.g., Montgomery & O'Donoghue 1999).

The "bootstrap" is an ideal way to assess the uncertainties in fitted parameters for a time-series analysis. The procedure is outlined in Wall & Jenkins (2003) and has recently been used in a number of *MOST* applications (e.g., Rowe et al. 2006b and Saio et al. 2006). Clement et al. (1992) also used a bootstrap to estimate the uncertainties in Fourier parameters they derived for

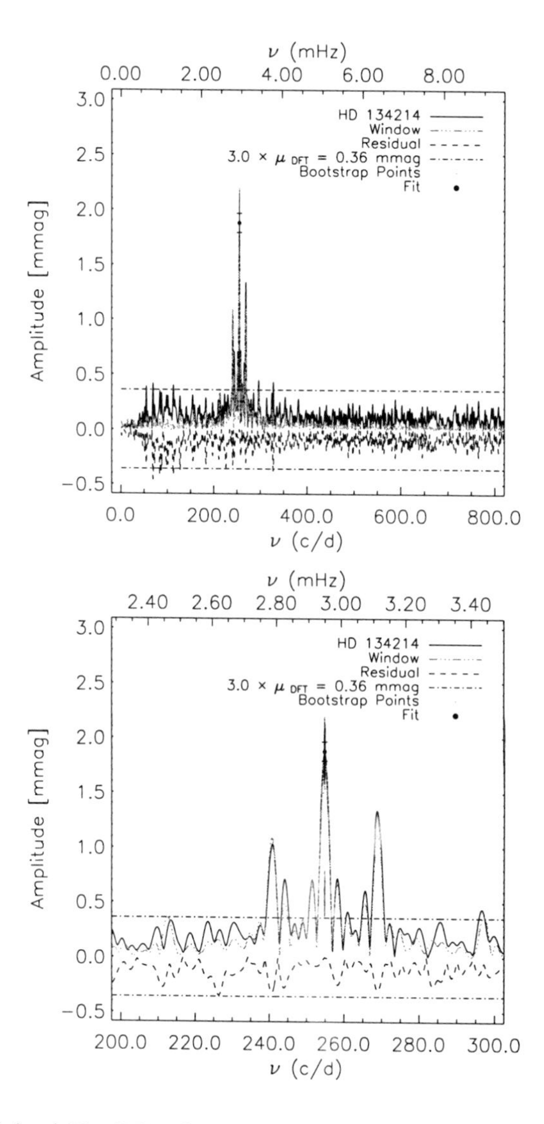


Figure 2: a) (top) The DFT of the HD 134214 data. Shown are the window function, the fit to the data (filled circle) and the bootstrap points (small dots). Also shown is the residual DFT after the fit; inverted for clarity. b) (bottom) The DFT of the data zoomed in on the region of the largest peak. Line styles are the same as in the top plot.

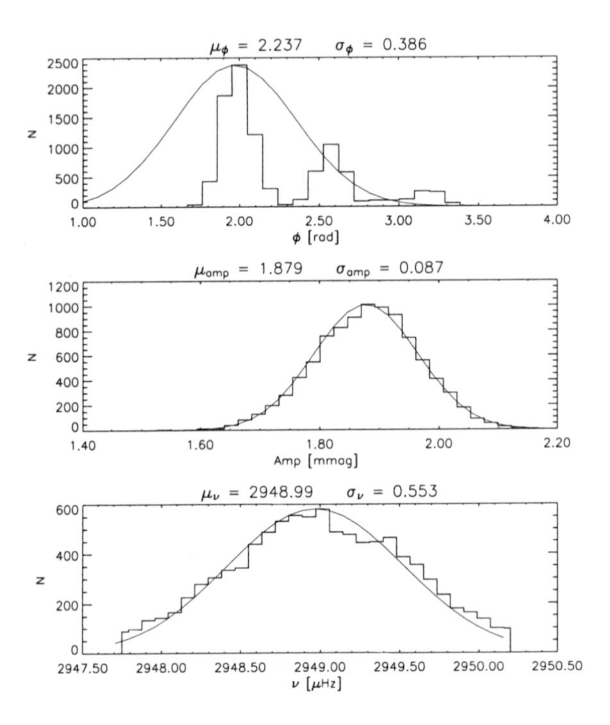


Figure 3: The bootstrap distributions for the fit parameters. Shown from top to bottom are the distributions for the phase $\phi$ in radians, the amplitude in mmag, and the frequency $\nu$ in $\mu$Hz. The labels above each panel give the mean $\mu$ and the standard deviation $\sigma$ calculated for each distribution. Gaussians with width $\sigma$ and centered on the original fit parameters are plotted over each distribution.

RR Lyrae stars.

In short, the bootstrap is a technique that allows a user to produce a distribution for each calculated parameter by constructing a large number of light curves from the original data. No assumptions need to be made about noise properties of the data and individual photometric errors are not needed. Each new light curve is assembled by randomly selecting N points from the original light curve (also containing N points) with the possibility of replacement. In this way, the new synthetic light curves preserve the noise properties of the original data. The fit procedure is repeated for each new light curve, eventually building distributions in each of the fit parameters. We then estimate the $1\sigma$ error bars from the analytic expression for the standard deviation of each distribution under the assumption that they are normally distributed. Each distribution is checked to ensure this assumption is valid.

Figure 3 shows the bootstrap distributions for 10,000 realisations of the fit to our data. The phase is the most uncertain parameter. Amplitude and frequency distributions are normal in shape. Overplotted are Gaussians with the standard deviations derived from the distributions. The Gaussians are centred on the original fit parameters. It should be stressed that bootstrapping only estimates the uncertainties in parameters and does not refine them or assign significances to them.

## Discussion

We determine that HD 134214 is monoperiodic and pulsating with a frequency of 2948.97 $\pm$ 0.55 $\mu$Hz. This is consistent with the earlier photometric observations of this star by Kreidl et al. (1994). Those authors detected frequency changes which they interpreted as cyclic with a timescale of about 1 year. The frequency resolution of our short $MOST$ run is insufficient to rigorously test this assertion. However, we have considered the implications of pulsational frequency variability in HD 134214.

Heller & Kawaler (1988) have suggested that it may be possible to exploit the frequency variability of roAp stars to determine their evolutionary status. However, the one-year timescale of variability in the case of HD 134214 is too short to be associated with evolution. We have also modelled the amplitude of the frequency change over the main sequence life time of an A-type star. We estimate that a 20 year baseline of observations would show a frequency change on the order of $10^{-4} - 10^{-5}$ $\mu$Hz. This is much smaller than the change of $\sim 0.2$ $\mu$Hz reported by Kreidl et al. (1994). An investigation by Cameron and co-workers has shown, that taking magnetic perturbations to the oscillation frequencies into account does not improve the discrepancy between the observed and calculated frequency changes.

Recent results by Kurtz et al. (2006) suggest that HD 134214 is oscillating spectroscopically in up to 6 modes. Our observations can rule out additional photometric oscillations at a $2\sigma$ detection level of 0.36 mmag. Our estimated DFT noise level of $\sim$ 0.12 mmag for 10 hours of observations with $MOST$ is comparable to that obtained by Kreidl et al. (1994) based on about 56 hours of photometry from four observatories during about 4 months. The excitation and selection of pulsation modes in roAp stars is an open question. In the case of HD 134214, there are several avenues to be explored with respect to the additional modes seen spectroscopically: (1) Is it a case of radial velocities of certain elements and ionisation stages being more sensitive to degrees of higher $\ell$? (2) If the sensitivity of the photometry can be improved sufficiently, will these modes also appear? (3) Is the broadband photometry just averaging over too wide an extent of the pulsating atmosphere? or (4) Might there be new physics in the upper atmospheres of (some) roAp stars to account for the differences observed?

We estimate that even observations over only a few days with *MOST* would (in addition to improving our frequency resolution enough to investigate cyclic variability) reduce our noise levels by approximately half. Even this modest improvement in noise would be valuable as a test of the Kurtz et al. (2006) theory that photometric observations are not sensitive to the new type of upper atmosphere pulsational variability they report.

**Acknowledgments.** JMM, DBG, AFJM, SR, and GAHW are supported by funding from the Natural Sciences and Engineering Research Council (NSERC) Canada. RK is funded by the Canadian Space Agency. WWW received financial support from the Austrian Science Promotion Agency (FFG - MOST) and the Austrian Science Funds (FWF - P17580).

## References

Audard, N., et al. 1998, A&A , 335, 954
Clement, C.M., Jankulak, M., & Simon, N. R. 1992, ApJ, 395, 192
Heller, C.H., & Kawaler, S.D. 1988, ApJ, 329, L43
Kreidl, T.J., et al. 1994, MNRAS , 270, 115
Kurtz, D.W. 1982, MNRAS , 200, 807
Kurtz, D.W. 1995, ASP Conf. Ser. 76: GONG 1994. Helio- and Astro-Seismology from the Earth and Space, 76, 60
Kurtz, D.W., & Martinez, P. 2000, Baltic Astronomy, 9, 253
Kurtz, D.W., Elkin, V.G., & Mathys, G. 2006, MNRAS , 370, 1274
Kuschnig, R., et al. 1997,A&A, 328, 544
Lenz, P., & Breger, M. 2005, CoAst, 146, 53
Matthews, J.M., et al. 2004, Nature, 430, 51
Montgomery, M.H., & O'Donoghue, D. 1999, DSSN, 13
Rowe, J.F., et al. 2006a, CoAST, 148, 34
Rowe, J.F., et al. 2006b, ApJ , 646, 1241
Saio, H., et al. 2006, ApJ, in press
Shavrina, A., et al. 2004, IAU Symposium, 224, 711
Sperl, M. 1998, CoAst, 111, 1
Walker, G., et al. 2003, PASP, 115, 1023
Walker, G., et al. 2005, ApJ, 635, L77
Wall, J.V., & Jenkins, C.R. 2003, Practical Statistics for Astronomers,Cambridge,Cambridge University Press

*Comm. in Asteroseismology*
*Vol. 148, 2006*

# Towards accurate component properties of the Hyades binary $\theta^2$ Tau

P. Lampens, Y. Frémat, P. De Cat and H. Hensberge

Koninklijke Sterrenwacht van België, Ringlaan 3, 1180 Brussel, Belgium

## Abstract

$\theta^2$ Tau is a well-detached, "single-lined" Hyades binary consisting of two mid-type A stars which both lie in the lower Cepheid instability strip. As a matter of fact, component A is a classical $\delta$ Scuti star showing a complex pattern of pulsations while component B might also be a $\delta$ Scuti pulsator (Breger et al. 2002). We acquired new high-resolution, high signal-to-noise spectra in order to determine accurate properties for the components of this system. Combining both spectroscopy and long-baseline optical interferometry, we were able to derive the orbital parallax and the component masses with unprecedented accuracy. Such constraints on the physical properties of the components are needed for a deep understanding of the pulsation physics. We also believe that $\theta^2$ Tau is an appropriate target to explore, in an empirical way, the possible interaction(s) between pulsation on the one hand and rotation and binarity on the other hand.

## Introduction

Our research currently focuses on binary and multiple stars with at least one pulsating component. In some cases, both the theories of stellar evolution and of pulsation can be tested and refined. Accurately derived component properties compared to suitably chosen theoretical isochrones allow to obtain information on the object's age and evolutionary status and to help discriminate among various possible pulsation models. We selected the $\delta$ Scuti star $\theta^2$ Tau for a

[1]Short research note based on a poster presented at IAU Symp. 240 "Binaries as Critical Tools and Tests in Contemporary Astrophysics", Prague, Aug. 2006

[2]Based on OHP (Observatoire de Haute-Provence) observations and spectral data retrieved from the ELODIE archive (http://atlas.obs-hp.fr/elodie/).

detailed study because a) it is a well-detached, spectroscopic binary resolved by long-baseline interferometry (Armstrong et al. 2006) it is a member of the Hyades open cluster at a mean distance of 45 pc (Perryman et al. 1998) the evolutionary status of its components is still under debate (Torres et al. 1997 (TSL97); Lastennet et al. 1999; Armstrong et al. 2006).

## Recent spectroscopic campaign: spectra and analysis

High-resolution spectra were acquired with the ELODIE Echelle spectrograph at the 1.93-m telescope of the Observatoire de Haute-Provence (OHP, France) through regular service mode observations from March 2005 until March 2006. These spectra were obtained over the full wavelength range (including the hydrogen lines) with a typical S/N of 100 and cover an entire orbital cycle. We also made use of some older observations from the ELODIE data base (Moultaka et al. 2004) as well as of the spectra acquired by TSL97 during the years 1989 till 1996 at the Oak Ridge Observatory (Harvard, Massachusetts).

Since the spectral lines of both components never separate completely due to the Doppler shifts being smaller than the line widths, we applied the spectra disentangling technique using the code KOREL developed by Hadrava (1995). For a description of the technique and usage, we respectively refer to Hadrava (2004) and Hensberge & Pavlovski (2007). Studied spectral regions were selected for their high intrinsic content of radial velocity information (Verschueren & David 1999, Hensberge et al. 2000). Application of KOREL to the above mentioned spectra enabled us to extract the individual contribution of each component to the observed spectra together with the orbital parameters, and therefore also to produce for each component of the binary a set of 117 KOREL radial velocities relative to the systemic velocity with a homogeneous coverage in amplitude and in orbital phase.

## Combined orbital analysis

We next combined the previous data set with 34 best-fit angular separations ($\rho$) and position angles ($\theta$) as derived from the interferometric measurements (Armstrong et al. 2006) . An astrometric-spectroscopic orbit was computed using the VBSB2 code which performs a global exploration of the parameter space followed by a simultaneous least-squares minimization (Pourbaix 1998). The combination of these measurement techniques is a powerful tool for obtaining accurate fundamental parameters. The resulting orbital elements and standard deviations of the best orbital solution in the sense of minimum least-squares residuals are presented in Table 1. The previous results of a similar computation performed by Torres et al. (1997) are also listed for comparison. While there is

Table 1: Orbital elements with standard deviations including orbital parallax and dynamical masses.

| Orbital element | This work | TSL97 results |
|---|---|---|
| P (days) | 140.7285 ± 0.0004 | 140.7282 ± 0.0009 |
| T | 1990.7630 ± 0.0002 | 1993.0752 ± 0.0008 |
| e | 0.7353 ± 0.0004 | 0.727 ± 0.005 |
| a (") | 0.0188 ± 0.0001 | 0.0186 ± 0.0002 |
| i (°) | 47.65 ± 0.12 | 46.2 ± 1.0 |
| Ω (°) | 354.59 ± 0.12 | 171.2 ± 1.8 |
| ω (°) | 234.61 ± 0.12 | 236.4 ± 1.1 |
| $V_0$ (km/s) | – | +39.5 ± 0.2 |
| $\kappa = \frac{M_B}{M_A+M_B}$ | 0.452 ± 0.002 | 0.46 ± 0.05 |
| $\pi_{dyn}$ (mas) | 21.20 ± 0.13 | 21.22 ± 0.76 |
| A (A.U.) | 0.8879 ± 0.0005 | 0.88 ± 0.04 |
| mass A ($M_\odot$) | 2.58 ± 0.04 | 2.4 ± 0.3 |
| mass B ($M_\odot$) | 2.13 ± 0.02 | 2.1 ± 0.2 |
| K1 (km/s) | 33.86 ± 0.11 | 33.18 ± 0.49 |
| K2 (km/s) | 40.98 ± 0.21 | 38 ± 2 |
| System mass ($M_\odot$) | 4.71 ± 0.10 (2.1%) | 4.54 ± 0.51 (11.2%) |
| Time span (yr) | 16.5 | 6.3 |

a good agreement for most orbital parameters, the most conspicuous difference is the larger radial velocity amplitude of component B. In particular, note the improvement in accuracy of the orbital parallax as well as on the dynamical component masses. Using this parallax together with $V$=3.40 ± 0.03 mag (Mermilliod et al. 1997) and Δm=1.12 ± 0.03 mag (Armstrong et al. 2006), we further derive the component absolute magnitudes $Mv_A = 0.36 \pm 0.04$ mag and $Mv_B = 1.48 \pm 0.04$ mag.

## Cluster membership and future work

Both components of $\theta^2$ Tau are among the more massive stars of the Hyades. They are located in the turnoff region of its colour-magnitude diagram. Provided that their physical properties are accurately known, both stars are useful indicators of chemical composition as well as age in this region (through fitting of theoretical isochrones). They also should allow to verify whether or not convective core overshooting occurs (Lebreton et al. 2001), as illustrated by the

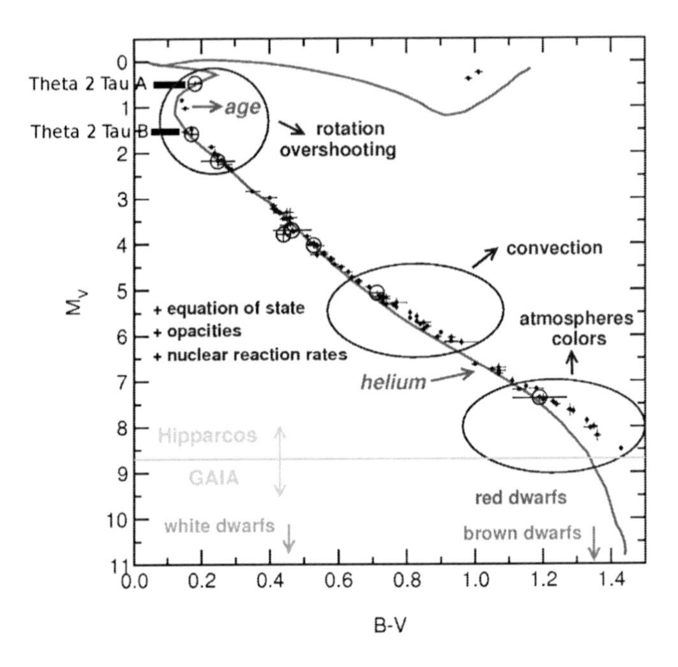


Figure 1: Hyades colour-magnitude diagram with an isochrone, 650 Myr age, Y=0.26, [Fe/H]=+0.14. and the locus of the components of $\theta^2$ Tau. (Courtesy of Y. Lebreton, GAIA Information Sheets)

Hyades colour-magnitude diagram in Fig. 1.

Furthermore, Breger et al. ( 2002) proposed that both components might be $\delta$ Scuti pulsators. Therefore, the knowledge of accurate fundamental component properties holds great potential for a reliable pulsation modelling of each star. Further work will consist in using the disentangled component spectra to perform a detailed chemical analysis, to determine as accurately as possible the physical properties and evolutionary status and to carefully test whether or not convective overshooting is needed in the models. If more high-resolution spectra could be obtained with a much higher temporal resolution, we also would be able to study the pulsation characteristics in the line profiles of $\theta^2$ Tau A and $\theta^2$ Tau B.

**Acknowledgments.** We acknowledge funding from the OPTICON Transnational Access Programme for spectra collected with ELODIE (OHP). We thank P. Hadrava and D. Pourbaix for supplying the codes KOREL and VBSB2 respectively. We further thank G.Torres for providing us the Oak Ridge observations. Financial support from the Belgian Federal Science Policy is gratefully acknowledged (projects IAP P5/36 and MO/33/018).

## References

Armstrong, J.T., et al. 2006, AJ, 131, 2643
Breger, M., et al. 2002, MNRAS, 336, 249
Hadrava, P. 1995, AAS 114, 393
Hadrava, P. 2004, in: "Spectroscopically and Spatially Resolving the Components of Close Binary Stars", Hilditch, R.W., Hensberge, H., & Pavlovski, K. (eds.), ASP Conf. Ser., 318, 86
Hensberge, H., Pavlovski, K., & Verschueren, W. 2000, A&A, 358, 553
Hensberge, H., & Pavlovski, K. 2007, Proc. of IAU Symp. 240 "Binaries as Critical Tools and Tests in Contemporary Astrophysics", Prague, Aug. 2006, in press
Lastennet, E., et al. 1999, A&A, 349, 485
Lebreton, Y., et al. 2001, A&A, 374, 540
Mermilliod J.C., Hauck B., & Mermilliod M. 1997, A&AS 124, 349
*http://obswww.unige.ch/gcpd/gcpd.html*
Moultaka, J., et al. 2004, PASP, 116, 693
Perryman, M., et al. 1998, A&A, 331, 81
Pourbaix, D. 1998, AAS, 131, 377
Torres, G., et al. 1997, ApJ, 485, 167
Verschueren, W., & David, M. 1999, A&AS, 136,591

*Comm. in Asteroseismology*
*Vol. 148, 2006*

# New frequency analyses of the multiperiodic $\delta$ Scuti variables CR Lyn and GG UMa

C.W. Robertson [1], P. Van Cauteren[2], P. Lampens[3], E. García-Melendo[4], R. Groenendaels[5], J. Fox[6], P. Wils[7]

[1] SETEC Observatory, Goddard, KS, USA, (cwr@pixius.net)
[2] Beersel Hills Observatory (BHO), Laarheidestraat 166, B–1650 Beersel, Belgium (paulvancauteren@skynet.be)
[3] Koninklijke Sterrenwacht van België, B–1180 Brussel, Belgium (patricia.lampens@oma.be)
[4] Esteve Duran Observatory (EDO), Montseny 46 – Urb. El Montanya, 08553 Seva, Spain (duranobs@astrogea.org)
[5] Dworp Observatory, B-1653 Dworp, Belgium (roger.groenendaels@skynet.be)
[6] Summer fellow, Koninklijke Sterrenwacht van België
[7] Vereniging Voor Sterrenkunde, Belgium (patrick.wils@cronos.be)

## Abstract

CCD photometric time-series observations of the $\delta$ Scuti stars CR Lyn and GG UMa reveal that both are multiperiodic pulsators, with at least three close, independent frequencies for CR Lyn and two frequencies in the case of GG UMa. In addition SAO 14823, another star in the field of GG UMa, is reported to be new multiperiodic variable as well.

## Introduction

The Hipparcos mission discovered a large number of new variable stars including various $\delta$ Scuti stars (ESA 1997). Koen (2001) performed a statistical analysis of the HIPPARCOS epoch photometry in search of multiperiodicity among all stars which were classified as variable in the HIPPARCOS catalogue: periodic, unsolved and microvariables. Two of the multiperiodic stars detected among the periodic variables by Koen (2001), CR Lyn (= HIP 40651) and GG UMa (= HIP 45693), have been re-analysed using recent CCD time-series data collected at (mostly) private small observatories. Table 1 gives an overview of the instruments and photometry reduction packages used.

Table 1: Instruments used for this study.

| Observer(s) | Observatory | Telescope | CCD camera | Software |
|---|---|---|---|---|
| PVC, PL | BHO | 40-cm Newt. | SBIG ST10XMe | Mira AP |
| PVC | BHO | 25-cm Newt. | SBIG ST10XMe | Mira AP |
| PVC | Hoher List | 12.5-cm refr. | SBIG ST10XMe | Mira AP |
| CWR | SETEC | 30-cm Cass. | SBIG ST-8i | AIP4Win |
| EGM | EDO | 60-cm Cass. | SX Starlight | LAIA |
| RG | Dworp | 30-cm Newt. | Hisis44 | AIP4Win |

## CR Lyn

In 2002 and 2003, we collected a total of 9348 $V$-measurements for CR Lyn during 254.4 hours in 42 nights. The observational log of the data is presented in Table 2. The last column (RMS) lists the nightly range of the variation in differential magnitude between the two comparison stars in the field. These are HIP 40632 ($V = 8.73$, $B - V = 0.48$) and HIP 40562 ($V = 8.91$, $B - V = 0.75$). From the $B$ and $V$-data acquired at two observatories, a colour index of $B - V = 0.35$ at maximum and $B - V = 0.38$ at minimum was derived. Two individual light curves are illustrated in Fig. 1, from which the cycle-to-cycle variation of the amplitude is clearly seen. The maximum total $V$-amplitude is about 0.13 mag.

CR Lyn (= HIP 40651, V=7.65 mag, spectral type F0) was found by Koen (2001) to show frequencies of 7.59057 and 7.29149 c/d and amplitudes of 0.031 and 0.021 mag, respectively, from the HIPPARCOS photometry (see Table 2 in his work). A frequency analysis of the new $V$-data set using Period04 (Lenz & Breger 2005) shows the same two frequencies. However, in contrast to Koen's results, we found two equally dominant frequencies (i.e., within their probable errors). This can be seen in Table 3, where we list the frequencies and the linear combinations of these frequencies, together with their semi-amplitudes and phases, in order of decreasing amplitude. The signal-to-noise ratio was calculated after prewhitening for all the frequencies. Formal uncertainties are given between parentheses. The last column indicates the reduction in total variance accounted for by this frequency.

Since the amplitude in the $H_p$ passband is not exactly the same as the amplitude in the $V$ passband ($H_p$ broader), a direct comparison cannot be made. However, the ratio of both amplitudes can be compared and we suggest that either, the amplitude of one of the major frequencies changed over several years, or, an additional frequency is affecting the observed amplitude ratio. We

Table 2: Observational log for CR Lyn.

| Observers | Filter | Timespan (JD-2450000) | Nr of nights | Nr of hours | Nr of data | RMS (mag) |
|---|---|---|---|---|---|---|
| RG | $V$ | 2257-2342 | 8 | 37.0 | 2082 | 0.009-0.023 |
| CWR | $V$ | 2671-2739 | 11 | 73.9 | 1882 | 0.009-0.014 |
| PVC, PL | $V$ | 2681-2745 | 12 | 65.7 | 3239 | 0.005-0.013 |
| EGM | $V$ | 2709-2753 | 12 | 77.8 | 2145 | 0.005-0.009 |
| PVC, PL | $B$ | 2681-2689 | 4 | 24.6 | 352 | 0.007-0.013 |
| CWR | $B$ | 2708-2739 | 6 | 34.6 | 1930 | 0.008-0.022 |
| PVC, PL | $R$ | 2681-2689 | 3 | 15.9 | 290 | 0.007-0.015 |
| CWR | $R$ | 2708-2739 | 6 | 34.7 | 1934 | 0.008-0.020 |

Table 3: Observed frequencies for CR Lyn ($V$-data).

| Id. | Freq. (c/d) | Semi-ampl. (mag) | Phase (cycles) | S/N | Resid. (mag) | Reduct. var. |
|---|---|---|---|---|---|---|
| $f_1$ | 7.29172(1) | 0.0225(2) | 0.628(2) | 28.8 | 0.020 | 24% |
| $f_2$ | 7.59056(1) | 0.0196(2) | 0.398(2) | 25.5 | 0.016 | 39% |
| $f_3$ | 7.31341(2) | 0.0104(2) | 1.000(3) | 13.3 | 0.015 | 45% |
| $f_1 + f_2$ | 14.88228(2) | 0.0024(2) | 0.295(14) | 5.5 | 0.014 | 46% |
| $2f_2$ | 15.18112(2) | 0.0023(2) | 0.242(15) | 5.2 | 0.014 | 46% |
| $f_2 + f_3$ | 14.90397(3) | 0.0018(2) | 0.787(19) | 4.1 | 0.014 | 46% |

prewhitened the data for the two main frequencies and searched for yet another one. Indeed, a third frequency, $f_3$, was detected very close to $f_1$ in the power spectrum obtained after prewhitening (Fig. 2). The day$^{-1}$ aliases of $f_3$ are also visible. The difference between $f_3$ and $f_1$ is only 0.02 c/d, but well above the theoretical resolution of $1.5/\Delta T$ = 0.003 c/d for our data set.
Because of the much smaller time span, no frequency searches were performed using the $B$ and $R$-data. Instead, amplitudes and phases were calculated for the frequencies we derived from the $V$-data set. These are listed in Table 4.

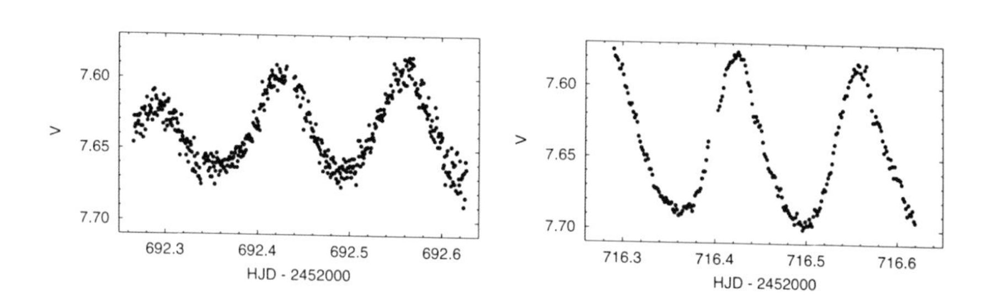


Figure 1: $V$ light curves of CR Lyn for two nights. Data are from BHO (left) and EDO (right panel).

Table 4: Semi-amplitudes and phases for CR Lyn in the filters $B$ and $R$.

| Id | Semi-ampl. $B$ (mag) | Phase $B$ (cycles) | Semi-ampl. $R$ (mag) | Phase $R$ (cycles) |
|---|---|---|---|---|
| $f_1$ | 0.0336(6) | 0.330(3) | 0.0249(5) | 0.748(4) |
| $f_2$ | 0.0265(6) | 0.476(4) | 0.0194(5) | 0.662(5) |
| $f_3$ | 0.0126(7) | 0.166(8) | 0.0104(6) | 0.200(10) |
| $f_1 + f_2$ | 0.0019(5) | 0.203(43) | 0.0024(5) | 0.855(32) |
| $2f_2$ | 0.0013(6) | 0.318(67) | 0.0020(5) | 0.637(44) |
| $f_2 + f_3$ | 0.0015(6) | 0.205(60) | 0.0013(6) | 0.371(73) |

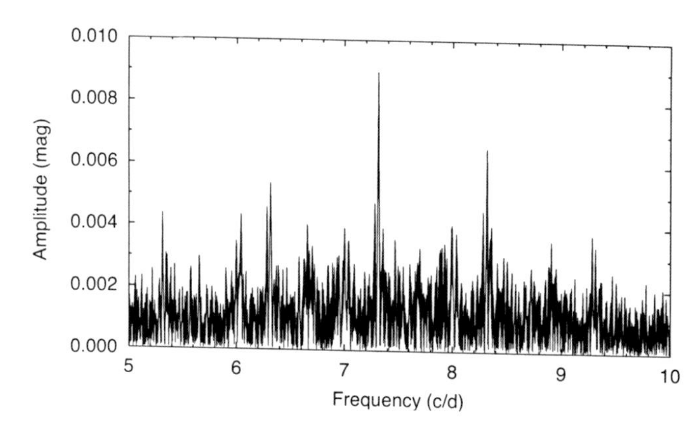


Figure 2: Frequency spectrum of CR Lyn after prewhitening $f_1$ and $f_2$.

## GG UMa

We collected a total of 3385 $V$-measurements for GG UMa during 75.2 hours in 15 nights during the winter 2003-2004. The observational log is presented

Table 5: Observational log for GG UMa.

| Observers | Filter | Timespan (JD-2450000) | No of nights | No of hours | No of data | RMS (mag) |
|---|---|---|---|---|---|---|
| PVC, PL | $V$ | 2992-3120 | 9 | 38.1 | 2258 | 0.016-0.030 |
| CWR | $V$ | 3018-3095 | 6 | 37.0 | 1960 | 0.011-0.029 |
| PVC, PL | $B$ | 3095-3097 | 2 | 9.4 | 212 | 0.020-0.025 |

Table 6: Observed frequencies for GG UMa.

| Id. | Freq. (c/d) | Semi-ampl. (mag) | Phase (cycles) | S/N | Residual (mag) | Reduction var. |
|---|---|---|---|---|---|---|
| $f_1$ | 7.41526(6) | 0.0319(4) | 0.108(2) | 18.8 | 0.021 | 36% |
| $f_2$ | 7.83527(11) | 0.0179(4) | 0.969(4) | 10.4 | 0.017 | 48% |

in Table 5. The $B$-filter data collected at BHO were not used in the frequency analysis. They were used to derive an average $B - V$ for GG UMa of 0.43. The following comparison stars were used: SAO 14841 (= GSC 4138-0077, $V = 9.52$, $B - V = 0.61$) and GSC 4138-0061 ($V = 11.65$, $B - V = 0.91$). Two light curves are shown in Fig. 3. Koen (2001) derived two frequencies, 7.41615 and 7.83449 c/d, with the respective amplitudes of 0.029 and 0.018 mag for GG UMa (= HIP 45693, V=8.60 mag, spectral type F5) from the HIPPARCOS mission. We confirm both from the frequency analysis of our $V$-data set. The frequencies with their corresponding semi-amplitudes and phases are listed in Table 6. As before, we also list the signal-to-noise ratio (after prewhitening), the residual RMS and the fraction of the reduction in total variance.

Although the residuals after prewhitening for both frequencies are still large (0.017 mag; note that because the check star was relatively faint, the standard deviation of its differential magnitude is larger as well), none of the additional candidate frequencies found so far had a signal-to-noise ratio higher than 4. Therefore, they may be spurious. We conclude that, given the much shorter time span of this $V$-data set (only 128 days), any additional frequency cannot be resolved at this stage. GG UMa may be an evolved star as the F5 spectral type is cool for a main-sequence $\delta$ Scuti variable.

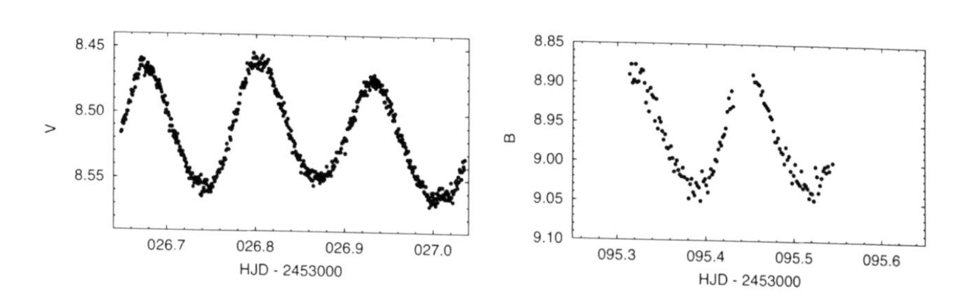


Figure 3: Light curves of GG UMa for two nights. $V$-data are from SETEC (left) and $B$-data from BHO (right panel).

## SAO 14823

We further report the detection of short-period variability in the light curves of SAO 14823 (= GSC 4138-1569), which is located in the same field as GG UMa (see Fig. 4). Our data reveal a frequency of 23.7 c/d with a total amplitude in $V$ of almost 0.04 mag as well as slow variations on a timescale of several days and with similar amplitude. Being of spectral type A0, this star may well be another $\delta$ Scuti star. Another phenomenon than pulsation might be needed to explain the existence of the slow variations found. However, our data are presently quite insufficient to characterise this longer-term variability.

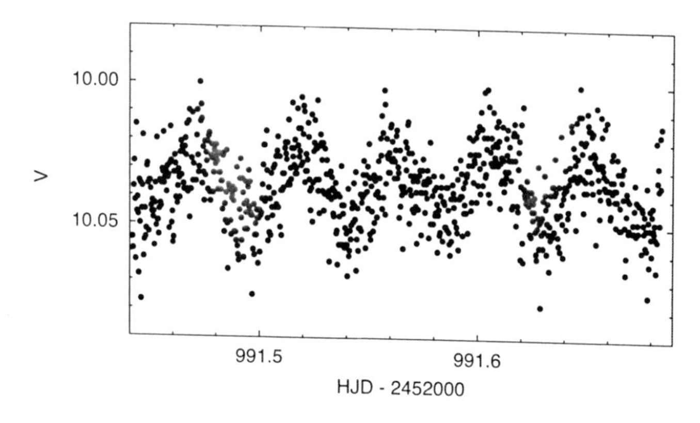


Figure 4: Light curve of SAO 14823 from BHO.

## Conclusions

New time-series CCD photometry was acquired for two $\delta$ Scuti variable stars found to be multiperidiodic by Koen (2001). In the case of CR Lyn, the total time span of our data equals almost 500 days and we were able to identify three independent frequencies, two of which ($f_1$ and $f_3$) are very close to each other. In addition, an inverted amplitude ratio for the frequencies $f_1$ and $f_2$ was detected with respect to Koen's (2001) findings based on HIPPARCOS data. In the case of GG UMa, we confirm the previous frequency analysis and are unable to detect more than two frequencies, mainly due to the much shorter timespan of our data set (based on one season only) We also report the detection of very short-period variations as well as a longer-term variability in the light curves of SAO 14823, located in the field of GG UMa.

**Acknowledgments.** The Belgian data have been acquired with equipment purchased thanks to a research fund financed by the Belgian National Lottery (1999). EGM thanks Joan A. Cano and Rafael Barberá for developing the LAIA software. PL acknowledges support from the Fund for Scientific Research (FWO) - Flanders (Belgium) through project G.0178.02. PVC thanks Dr. K. Reif of the Hoher List Observatory, Universitätsternwarte Bonn, for the allocated telescope time. JF acknowledges a fellowship from the Belgian Science Policy Office during August 2004. This research made use of the Simbad and VizieR databases operated by the *Centre de Données Astronomiques*, Strasbourg, France.

## References

Koen, C. 2001, MNRAS, 321, 44
Lenz, P., & Breger, M. 2005, CoAst, 146, 53
ESA 1997, The HIPPARCOS and TYCHO catalogues, ESA SP–1200

*Comm. in Asteroseismology*
*Vol. 148, 2006*

# Spectroscopy of HIP 113790 at the 5th OHP NEON summer school

Y. Frémat[1], A. Antonova[2], Y. Damerdji[3], C.J. Hansen[4], M.T. Lederer[5], M. Tüysüz[6], P. Lampens[1], P. Van Cauteren[7]

[1] Royal Observatory of Belgium, 3 avenue circulaire, 1180 Brussels, Belgium
[2] Armagh Observatory, College Hill, Armagh, BT61 9DG, N. Ireland
[3] Observatoire de Haute-Provence, 04870 St Michel l'Observatoire, France
[4] N. Bohr Institute, Astronomy, Juliane Maries Vej 30, 2100 Copenhagen, Denmark
[5] Institut für Astronomie, Türkenschanzstrasse 17, 1180 Wien, Austria
[6] Department of Physics, Faculty of Arts and Sciences, Çanakkale Onsekiz Mart University, 17100, Çanakkale, Turkey
[7] Beersel Hills Observatory (BHO), 1650 Beersel, Belgium

## Abstract

In this contribution, we present the high–resolution observations we obtained for HIP 113790 at the Haute-Provence Observatory using the AURELIE spectrograph. By determing its fundamental parameters, we confirm that the star is located in the $\delta$ Scuti instability strip. We further report the detection of rapid line profile variations with a period of about 1.43 hours.

## Introduction and Observations

HIP 113790 is a bright (V=7.30) poorly studied $\delta$ Scuti star (Frémat et al. 2006a) that shows radial velocity variations (Grenier et al. 1999) and that was classified "unsolved variable" in the HIPPARCOS catalogue (ESA 1997). Preliminary photometric observations (Frémat et al. 2005) were carried out in December 2004 and showed a complex multiperiodic variation pattern with a main period of about 1.2 hours. Taking advantage of the observing time available during the 5$^{th}$ NEON school at the Observatoire de Haute Provence (OHP), we obtained time series of high-resolution spectra for 3 nights on the 1.52 m telescope equipped with the AURELIE spectrograph (Gillet et al. 1994). The observations log is given in Table 1 with Date (col.1), grating number (col.2), spectral range (col.3), resolution (col.4), time exposure (col.5) and number of exposures (col.6). Our aim was twofold: to derive the stellar parameters of the star and to interpret the radial velocity variability in terms of line profile variations.

Table 1: Description of the spectroscopic observations.

| Date | Grating | Spectral range | R | t (s) | n |
|---|---|---|---|---|---|
| July 26, 2006 | 3 | 4000 – 4430 | 5500 | 1200 | 5 |
| July 28, 2006 | 7 | 5380 – 5520 | 28000 | 1200 | 14 |
| July 30, 2006 | 7 | 5380 – 5520 | 28000 | 1200 | 12 |

## Stellar parameters

The stellar parameters of HIP 113790 were derived by comparing the observed 4000 – 4430 Å wavelength range to synthetic spectra. Model atmosperes we used were computed by Castelli & Kurucz (2003) with the ATLAS9 program and the theoretical spectra were obtained with the SYNSPEC (Hubeny & Lanz 1995, see references therin) computer code for a solar-like chemical composition and a 2 km s$^{-1}$ microturbulent velocity. A least-squares algorithm (Frémat et al. 2006b) was used to derive, first, the projected rotation velocity ($V\sin i$) and the heliocentric radial velocity (RV) by fitting several metallic lines, then the effective temperature (T$_{\rm eff}$) and surface gravity (log $g$) by studying the H$\gamma$ and H$\delta$ line profiles. Our results are as follows:

$$\begin{aligned} V\sin i &= 28 \pm 3 \text{ km s}^{-1} \\ \text{RV} &= 5.3 \pm 0.74 \text{ km s}^{-1} \\ \text{T}_{\rm eff} &= 7185 \pm 100\ K \\ \log\ g &= 3.80 \pm 0.12. \end{aligned}$$

The averaged radial velocity we obtain is similar to the value (RV = 5.8 $\pm$ 7.6 km s$^{-1}$) published by Grenier et al. (1999), and there is therefore no sign of multiplicity. We further noticed no chemical peculiarity in the spectra of HIP 113790 and we obtained a very good agreement between observations and theory as shown in Fig. 1. However, since the effective temperature of the star is close to the limit at which the hydrogen lines become less sensitive to surface gravity, we also estimated its log $g_\pi$ by computing the luminosity, from the V magnitude and the trigonometric parallax, and by interpolating the stellar mass in theoretical evolutionary tracks (Schaller et al. 1992). The value we obtained (log $g_\pi$ = 3.89$\pm$0.09) is in fair agreement with the value derived from spectroscopy.

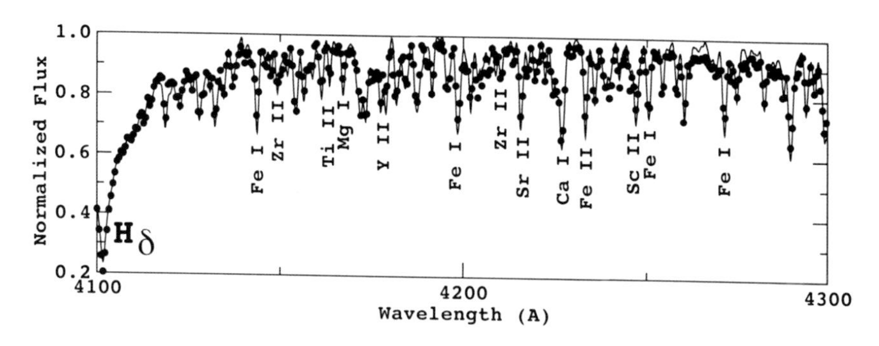


Figure 1: Comparison between observations (full line) and synthetic spectrum (dots).

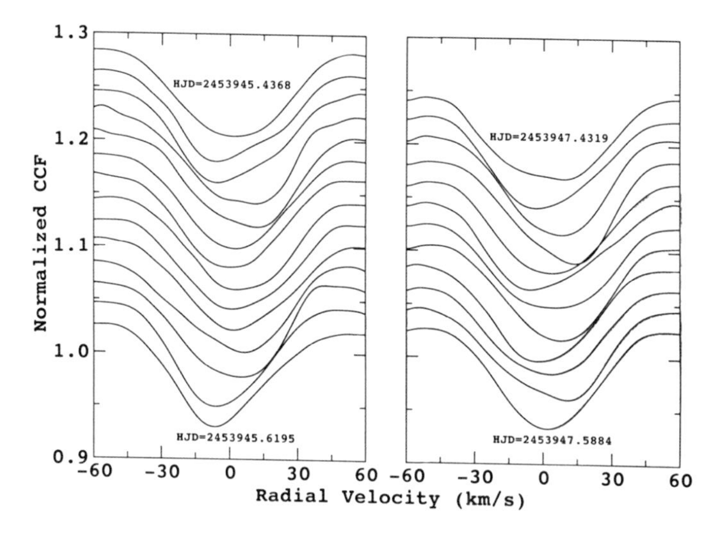


Figure 2: Variation of the cross-correlation function during two nights.

## Variability: Spectroscopy

We performed a preliminary study of the spectroscopic variations of HIP 113790 in a spectral domain ranging from 5380 to 5520 Å. In order to increase the S/N ratio of the observations, each exposure was cross-correlated with a synthetic spectrum computed adopting the $T_{eff}$ and the log $g$ values we previously derived. The time series of the resulting cross-correlation functions (CCF) are plotted in Fig. 2 for two nights and clearly show rapid spectroscopic variations with a period of about 1.43 hours, as we determined from the study of the residuals at the line center using the PERIOD04 software (Lenz & Breger 2005).

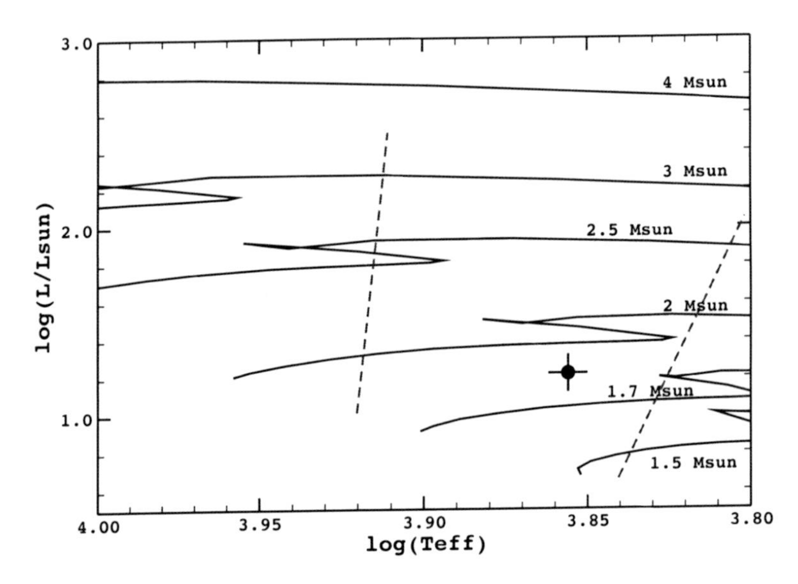


Figure 3: Location of HIP 113790 in the HR diagram. Broken lines represent the expected borders of the $\delta$ Scuti instability strip (Dupret et al. 2005). Evolutionary tracks (full lines) are from Schaller et al. (1992).

## Conclusions

As shown by the stellar parameters we derived, HIP 113790 is located in the $\delta$ Scuti instability strip (Fig. 3). From the study of our cross-correlation function time series, the radial velocity variability can definitively be interpreted as due to periodic line profile variations, with a main period (i.e. 1.4 hours) of the same order as the variations detected in the photometry (i.e. 1.2 hours). Additional spectroscopic observations would however be required on a larger telescope to perfom a complete frequency analysis on several individual spectral lines. During the past year, an extensive photometric follow-up of HIP 113790 was further carried out using the V and B filters at different observatories in Belgium (P.Lampens & P.Van Cauteren), in Greece (K.D.Gazeas & P.G.Niarchos, S.Klides) and in the USA (C.W.Robertson). A full analysis of these data is presently undertaken.

**Acknowledgments.** We thank the OPTICON programme, M.Dennefeld and D.Gillet for funding our participation to the NEON summer school. Financial support from the Belgian Federal Science Policy is gratefully acknowledged (projects IAP P5/36 and MO/33/018).

## References

Castelli, F., & Kurucz, R. 2003, IAU Symposium 210, 20
ESA 1997, The Hipparcos and Tycho Catalogues ESA-SP 1200
Frémat, Y., Lampens, P., Van Cauteren, P., & Robertson, C.W. 2005, CoAst, 146, 6
Frémat, Y., Lampens, P., Van Cauteren, P., & Robertson, C.W. 2006a, Mm.S.A.It., 77, 174
Frémat, Y., Neiner, C., Hubert, A.M., et al. 2006b, A&A, 451, 1053
Gillet, D., Burnage, R., Kohler, D., et al. 1994, A&AS, 108, 181
Grenier, S., Baylac M.O., Rolland L., et al. 1999, A&AS, 137, 451
Hubeny, I., & Lanz, T. 1995, ApJ, 439, 875
Lenz, P., & Breger, M. 2005, CoAst, 146, 53
Schaller, G., Schaerer, D., Meynet, G., & Maeder, A. 1992, A&AS, 96, 269